Instabilities and Fronts
in Extended Systems

Princeton Series in Physics

edited by Philip W. Anderson, Arthur S. Wightman, and Sam B. Treiman

Quantum Mechanics for Hamiltonians Defined as Quadratic Forms *by Barry Simon*

Lectures on Current Algebra and Its Applications *by Sam B. Treiman, Roman Jackiw, and David J. Gross*

Physical Cosmology *by P.J.E. Peebles*

The Many-Worlds Interpretation of Quantum Mechanics *edited by B. S. DeWitt and N. Graham*

Homogeneous Relativistic Cosmologies *by Michael P. Ryan, Jr., and Lawrence C. Shepley*

The $P(\phi)_2$ Euclidean (Quantum) Field Theory *by Barry Simon*

Studies in Mathematical Physics: *Essays in Honor of Valentine Bargmann edited by Elliott H. Lieb., B. Simon, and A. S. Wightman*

Convexity in the Theory of Lattice Gases *by Robert B. Israel*

Works on the Foundations of Statistical Physics *by N. S. Krylov*

Surprises in Theoretical Physics *by Rudolf Peierls*

The Large-Scale Structure of the Universe *by P. J. E. Peebles*

Statistical Physics and the Atomic Theory of Matter, From Boyle and Newton to Landau and Onsager *by Stephen G. Brush*

Quantum Theory and Measurement *edited by John Archibald Wheeler and Wojciech Hubert Zurek*

Current Algebra and Anomalies *by Sam B. Treiman, Roman Jackiw, Bruno Zumino, and Edward Witten*

Quantum Fluctuations *by E. Nelson*

Spin Glasses and Other Frustrated Systems *by Debashish Chowdhury* (*Spin Glasses and Other Frustrated Systems* is published in co-operation with World Scientific Publishing Co. Pte. Ltd., Singapore.)

Large-Scale Motions in the Universe: A Vatican Study Week *edited by Vera C. Rubin and George V. Coyne, S.J.*

Instabilities and Fronts in Extended Systems *by Pierre Collet and Jean-Pierre Eckmann*

Instabilities and Fronts in Extended Systems

Pierre Collet

Jean-Pierre Eckmann

Princeton University Press

Princeton, New Jersey

Published by Princeton University Press,
41 William Street, Princeton, New Jersey 08540
In the United Kingdom:
Princeton University Press, Oxford

Library of Congress Cataloging-in-Publication Data
Collet, Pierre, 1948–
Instabilities and fronts in extended systems /
P. Collet, J.-P. Eckmann.
p. cm.—(Princeton series in physics)
ISBN 978-0-691-63617-7
1. Bifurcation theory. 2. Stability. 3. Differentiable dynamical systems.
I. Eckmann, Jean-Pierre. II. Title. III. Series. QA374.C59 1990
515′.353—dc20 89–24050

This book has been composed in Linotron Times Roman

Princeton University Press books are printed on acid-free paper,
and meet the guidelines for permanence and durability of the
Committee on Production Guidelines for Book Longevity
of the Council on Library Resources

Printed in the United States of America
by Princeton University Press,
Princeton, New Jersey
10 9 8 7 6 5 4 3 2 1

Pierre Collet
Centre de Physique Théorique
Laboratoire UPR A0014 du CNRS
Ecole Polytechnique
Route de Saclay
F 91128 Palaiseau CEDEX (France)

Jean-Pierre Eckmann
Département de Physique Théorique
et Section de Mathematiques
Université de Genève
24, quai Ernest-Ansermet
CH 1211 Genève 4 (Suisse)

Contents

Preface

The physics of extended systems is a subject of great interest for the Experimentalist and the Theoretician alike. There exists an enormous literature on this subject in which solutions, bifurcations, fronts, and the dynamical stability of these objects are discussed. To the uninitiated reader, the theoretical methods which lead to the various results seem often somewhat *ad hoc* and it is not clear how to generalize them to the next – that is, not yet solved – problem.

The aim of this book is to give a more systematic account of these methods, and to work out the relevant features which make them operational. As far as possible, we shall give proofs for the applicability of a method in the following situations: We formulate conditions which are sufficiently general to cover at least some of the relevant equations found in the literature. On the other hand we – on purpose – do not strive for the most general formulation for which we can imagine that a given method works, because this would probably obscure the intrinsic simplicity of a method.

It should be kept in mind that the subject of instabilities in an infinite domain is intrinsically difficult, in particular because of the appearance of problems related to the continuous spectrum. Many seemingly innocent questions have, to date, not even the beginning of a satisfactory mathematical answer; most often it is in fact not known what the correct questions are. We will, on occasion, point out what seems, to us, to be the way in which further problems could go, and indicate possible methods which could lead along that way. However, it is our belief, that such progress is made foremost by studying new, and more complicated examples. We hope that this book can serve as a guide to what can be considered "well-known" or at least "solved" in this subject and to what are new and open problems.

Some of the material in this book is well-documented in the literature. Rather than referring to the literature, we have preferred to give a presentation of most of these subjects except those we think should form the basic background of any reader trying to understand our subject. These are the elementary theories of matrices and linear operators, eigenvalues and the like; some notions on differential equations; and some notions on analytic functions, such as the Cauchy formula. A glossary at the end of the book may help to review some of the slightly less common subjects.

Outline

Although this book contains a multitude of small sections, its design is governed by an overall plan. We shall now explain this plan. The set of physical systems described by hydrodynamics and such similar subjects has a rich phenomenology, incorporating such intricate details as vortices, weather forecasts and the like. To the mathematically inclined physicist, this richness is at the same time a challenge and a frustration. The challenge is a better understanding of these phenomena and the frustration comes from our awareness of how little is known. Of course, many scientists have important insights into how to attack hard problems of the above type. In this book, we pursue the line of studying somewhat less general, but hopefully more typical questions. This allows us to achieve greater precision in the statements of results and to work out methods which lead to proofs within the present knowledge of analysis.

We start with a study of the most simple physical situations – namely those where nothing moves, or where only simple motion takes place. As some external parameters of the problem are changed, such as a temperature gradient, the system may go into an excited state, and may show, for example, a new kind of oscillation. This transition from one state to another can be created by an *instability*. All developments in this book focus around this central notion. There are two major sets of instabilities which we study: those in a *finite dimensional state space* and those in an *infinite dimensional state space* with the additional requirement of *invariance of the physical system under a group of translations*. The book is organized accordingly.

- In Chapter I, we set the stage by explaining a few of the physical equations, introducing attractors, and pointing out informally the difference between finite and infinite systems.
- In Chapter II, we develop the theory of small solutions of nonlinear problems. This leads us naturally to the theory of invariant manifolds.
- In Chapter III, we start systematically with a study of bifurcation theory, mostly in the context of systems with a finite dimensional state space. We classify the most common bifurcations, and we show how this part of the theory can be useful even for infinite dimensional systems, when most of the infinitely many degrees of freedom are driven by finitely many "relevant" degrees of freedom. Finally, we state a very general bifurcation theorem, which is the basis for our further developments.
- In Chapter IV, we discuss stationary and quasistationary solutions obtained from constant solutions through bifurcation. We are working

in an infinitely extended domain, leading to a continuous spectrum for the possible excitations. Due to the translation invariance of the problem, we shall find that these bifurcation problems can still be handled in a simple way. We study in detail the linearized problem and provide existence proofs for stationary and convectively unstable solutions (solutions which bifurcate from a constant solution, but are only stationary in a moving frame).

- In Chapter V, we study the stability of the stationary solutions whose existence we established in Chapter IV. The stability will be seen to depend on the wavelength of the quasistationary solution ("Eckhaus instability"). We establish a relation between this instability and the existence of solutions to the equations in half-spaces. This can then be used to solve bifurcation problems in media which are *not translation invariant*. For example, this technique allows the study of bifurcations in a space where the physical parameter takes one value for negative coordinates and another for positive coordinates.
- In Chapter VI, we give a short account of the general ideas of multiscale analysis. In view of the scarcity of rigorous results, this chapter remains somewhat sketchy. However, the results of the following chapter are an implementation of these ideas.
- In Chapter VII, we explain some parts of the theory of fronts. Fronts can only occur in infinitely extended systems and correspond to solutions which are in one state at very negative coordinates and in another state at very positive coordinate values. For second order differential equations, a theory exists which is based on analysis of phase diagrams and on the maximum principle. For higher order equations, these methods are not available, and only multiscale analysis can be used.

Acknowledgments. During our occupation with the subject of this book, we have profited from enlightening discussions with many colleagues, in particular with J. Langer, who introduced us to the subject. Our collaboration was made possible by the support of the following institutions: University of Geneva, CNRS, Fonds National Suisse, IHES, NSF (Santa Barbara, Minneapolis), Rutgers University, Ecole de Physique de la Matière Condensée (Beg-Rohu). We thank the directors of these institutions who have made our stays agreeable and interesting.

Instabilities and Fronts
in Extended Systems

CHAPTER I
SETTING THE STAGE

1. Physical Equations

Partial differential equations describing the time evolution of a physical system are very common. They are derived either directly from basic physical principles or as some reduced descriptions of microscopical evolutions valid in some limit. We shall not discuss the question of establishing these macroscopic equations, this is a very difficult problem which is not completely understood.

To fix the ideas, we mention some of the most important macroscopic equations with the idea that the theories discussed below will apply without too much changes to all these equations. One of the best known equations is the Navier-Stokes equation which describes the time evolution of the velocity field $v(x,t) : \mathbf{R}^3 \times \mathbf{R} \to \mathbf{R}^3$ of a fluid. We denote by x an element of $\mathbf{R}^3$ and by x_i, $i = 1,2,3$ its components. It is given by

$$\partial_t v(x,t) = \nu \Delta_x v(x,t) - \big(v(x,t) \cdot \nabla_x\big) v(x,t) + f(x,t) - \nabla_x p(x,t) ,$$

with the incompressibility constraint

$$\nabla_x \cdot v(x,t) = 0 , \tag{1.1}$$

and where p is the pressure field, f the (bulk) force field, and ν the viscosity. The symbol $\cdot$ denotes the scalar product in $\mathbf{R}^3$. In finite volume one should add, of course, boundary conditions.

Other physical phenomena can take place in the fluid which are not described by the Navier-Stokes equation. For example, if the heat effects cannot be neglected one should couple the Navier-Stokes equation to the heat equation. Under some physical assumptions one gets the Boussinesq equation

$$\begin{aligned}
\partial_t v(x,t) &= \nu \Delta_x v(x,t) - \big(v(x,t) \cdot \nabla_x\big) v(x,t) \\
&\quad - \nabla_x p(x,t) + T(x,t) e_{x_3} , \\
\partial_t T(x,t) &= \kappa \Delta_x T(x,t) - v(x,t) \cdot \nabla_x T(x,t) , \\
\nabla_x \cdot v(x,t) &= 0 .
\end{aligned} \tag{1.2}$$

Here, e_{x_3} is a unit vector pointing in the vertical direction.

If the fluid is composed of charged particles one should add a coupling with the equation describing the evolution of an electromagnetic field and one obtains the so-called magneto-hydrodynamical system. We refer the reader to [Ch] for more information on these equations.

Similar equations also occur in the study of chemical systems. Assume a container is filled with N different chemical species. Each of them is described by its field of concentration $\rho_i(x,t) : \mathbf{R}^3 \times \mathbf{R} \to \mathbf{R}$ $(i = 1, \dots, N)$. The chemical evolution equations are given by

$$\partial_t \rho_i(x,t) + \nabla_x \cdot q_i(x,t) = F_i(x,t) ,$$

where $q_i : \mathbf{R}^3 \times \mathbf{R} \to \mathbf{R}^3$ is the flux and F_i is the source term for the species number i. The flux is most often given by

$$q_i(x,t) = -D_i \nabla_x \rho_i(x,t) + c_i \rho_i(x,t) ,$$

where the first term is a diffusion term (with diffusion constant D_i) and the second one is a transport term (at velocity $c_i \in \mathbf{R}^3$). The source term $F_i \in \mathbf{R}$ originates in the chemical reactions among the various species. It is usually a *nonlinear* function of the various concentrations.

Similar examples also occur in biology, combustion and the theory of flames, elasticity, metallurgy etc.

If these equations are considered in finite space volumes, one should add to them *boundary conditions*. These equations are also often considered in infinite volume, and we will later explain why. It turns out that these two problems are of a very different nature. The reason is that problems in finite volume can be reduced (at least formally) to finite systems of coupled differential equations, while this is not possible for infinite volume. In the latter case, the space variables can introduce new nontrivial effects, but we also get simplified formulations from the translation invariance.

2. Attractors

The time evolution of a physical system with finitely many degrees of freedom is often described by a system of differential equations of the form

$$dx/dt = X(x)$$

where the vector $x(t)$ represents the state of the system at time t. The set of all possible states is called the phase space of the system and is often a vector space $\mathbf{R}^d$. The vector valued function $X(x)$ is also called a vector field. The class of equations of the above type is called the class of dynamical systems. There are, of course, extensions of this formulation to vector fields on manifolds, or to partial differential equations but we shall first discuss the more elementary (but still very frequent and important) situation of differential equations.

We shall be interested mostly in the large time behavior of the evolution, and in particular in properties of this large time behavior which do not depend too much on the initial condition. While some transient properties are interesting and important in their own right, most questions in pure or applied physics deal with the large time behavior.

There are numerous examples of time evolution equations; to be specific, we shall illustrate some basic notions in the well-known case of Classical Mechanics. The evolution equations are Newton's equations and we shall discuss them in the Hamiltonian formalism. The phase space is even dimensional (of dimension $2d$) with a state represented by two d dimensional vectors q and p (respectively position and momentum). The evolution equations are constructed using a real valued function $H = H(p, q)$, called the Hamiltonian. They are given by

$$\frac{dq}{dt} = \frac{\partial H}{\partial p}, \quad \frac{dp}{dt} = -\frac{\partial H}{\partial q}.$$

A well-known and simple example is the harmonic oscillator with Hamiltonian given by $H = (p^2 + \omega^2 q^2)/2$ (the phase space dimension is 2). The quantity ω is constant. The evolution equations are easy to integrate, but since the value of the Hamiltonian is conserved during the time evolution, we conclude at once that the trajectories in the two dimensional phase space are ellipses, and the time evolution is simply going along these ellipses. This is the general situation for integrable Hamiltonian systems where the time evolution is taking place on invariant tori, cf. [Ar].

A slightly more interesting case is obtained if we add a linear friction term to the above time evolution. The system is no longer Hamiltonian, and not even conservative (energy is lost by friction). If the friction coefficient is denoted by the positive number η, the evolution equations are given by

$$\begin{cases} \dfrac{dq}{dt} = p \,, \\ \dfrac{dp}{dt} = -\omega^2 q - \eta p \,. \end{cases}$$

In Section 7 and in Section 19, we shall present in a more general context the method of replacing higher order differential equations by a first order system of equations. If we compute the time derivative of the Hamiltonian $H = (p^2 + \omega^2 q^2)/2$ for this evolution, we get

$$\frac{dH}{dt} = -\eta p^2 \,.$$

This implies the well-known fact that for any initial condition, the trajectory converges to the origin when time goes to infinity. We conclude that the origin, which is a stationary solution of the dynamics, describes the large time evolution of *any* initial condition.

This simple example also illustrates a common feature of dissipative dynamical systems which is the contraction of phase space. This implies that the large time dynamics is taking place on a smaller part of phase space. Intuitively, if we have a dissipative system, it will dissipate all its initial kinetic energy, and the asymptotic time evolution will be stationary as in the above example. In order to obtain a nontrivial asymptotic time evolution, we should therefore compensate the loss of energy by dissipation. This can be done by an external forcing. In what follows, we shall be concerned mostly with such dissipative systems. We want to emphasize, however, that many ideas and results can be adapted to the study of conservative systems.

Coming back to our simple example of the harmonic oscillator, we can now look at a damped oscillator with a harmonic forcing of period $2\pi/\alpha$ and amplitude A. The evolution equation is now given by

$$\begin{cases} \dfrac{dq}{dt} = p \,, \\ \dfrac{dp}{dt} = -\omega^2 q - \eta p + A\cos(\alpha t) \,. \end{cases}$$

The (well-known) solution of this system is given by

$$\begin{cases} q = \dfrac{A(\omega^2 - \alpha^2)}{(\omega^2 - \alpha^2)^2 + \alpha^2\eta^2} \cos\alpha t + \dfrac{A\alpha\eta}{(\omega^2 - \alpha^2)^2 + \alpha^2\eta^2} \sin\alpha t + \mathrm{tt} \,, \\ p = \dfrac{-A\alpha(\omega^2 - \alpha^2)}{(\omega^2 - \alpha^2)^2 + \alpha^2\eta^2} \sin\alpha t + \dfrac{A\alpha^2\eta}{(\omega^2 - \alpha^2)^2 + \alpha^2\eta^2} \cos\alpha t + \mathrm{tt} \,, \end{cases}$$

where "tt" means transitory terms which vanish as $t \to \infty$. We conclude immediately from these formulas that for any initial condition, the asymptotic time evolution is the same and is taking place on the same ellipse. By compensating the losses of energy due to the friction term we have obtained a nontrivial asymptotic dynamics. Note also that this asymptotic dynamics is stable in the sense that if we start with an initial condition which is near the ellipse, the trajectory will spiral towards the ellipse (in this simple example this is in fact true for any initial condition). The ellipse is sometimes called a dynamical stationary state or dynamical equilibrium. Note also that it depends on the parameters in the problem. If we change for example the friction coefficient, η, we get a different ellipse.

Summarizing the above examples, we see that the time evolution of a dissipative system contracts phase space. If the losses are not compensated, the evolution comes to rest. If they are compensated, a nontrivial dynamical equilibrium can appear in some part of phase space. The interesting dynamical equilibria are those which are locally attracting, that is, nearby initial conditions will give rise to trajectories which stay nearby and approach for large time the dynamical equilibrium (stability).

We can now formulate a mathematical definition for this notion of stable dynamical equilibrium.

Definition 2.1. *A set Ω is called an attractor for a vector field X if it is invariant and if there is a neighborhood V of Ω such that for every sufficiently small neighborhood U of Ω, and for large enough t, the flow S_t generated by X satisfies $S_t(V) \subset U$.*

Remark. One usually considers irreducible attractors. These are attractors containing a dense orbit.

We also want to emphasize that in the phase space of a dynamical system, several attractors can coexist in different regions. For a given attractor, the set of initial conditions which converge for large time to the attractor is called the basin (of attraction) of the attractor. In phase space, there can be other invariant sets, such as the boundary between basins of attraction, or repellers. A simple example with several attractors can be constructed as follows. Consider the Hamiltonian dynamical system in $\mathbf{R}^2$ associated with the Hamiltonian $H = p^2/2 + V(q)$, where the potential is given by $V(q) = -q^2/2 + q^4/4$. If we add a friction term as before, it is easy to verify that the system has two attractors, $(q = \pm 1, p = 0)$, and a stationary solution at the origin, $(q = 0, p = 0)$ which is not an attractor. Most initial conditions near the origin flow toward one or the other attractor (we shall give later a precise discussion of this local behavior).

The attractors depend, of course, on the dynamical system. In the above example with friction, consider the one parameter family of potentials $V_\lambda = -\lambda q^2/2 + q^4/4$. It is easy to show that for $\lambda < 0$ there is only one attractor at the origin. For $\lambda > 0$, the origin is no longer an attractor, but a repeller, and in addition there are two other attractors at $(q = \pm\sqrt{\lambda}, p = 0)$.

3. Finite and Infinite Space Systems

In this section we explain the main difference between systems in finite volume and systems in infinite volume.

We shall first discuss the ideas for the concrete example of the Rayleigh-Bénard experiment. This experiment is realized as follows. One considers a cubic container of size l which contains a fluid (in general water, oil, or regular liquid helium).

The vertical edges of the container are thermally insulated while the horizontal plates are good thermal conductors. One then imposes a temperature difference between the two horizontal plates, the top one being at a lower temperature. This experimental setup is described by the system of Boussinesq equations which gives the time evolution of the field of temperature $T(x,t)$ with the bottom plate at temperature 0 and of the velocity field of the fluid $v(x,t)$. In contrast to equation (1.2), we have subtracted from T a linear function such that the new T equals 0 at the bottom and 0 at the top. The evolution equation is given explicitly by

$$\begin{aligned} P^{-1}\left(\partial_t v + (v \cdot \nabla_x)v\right) &= \Delta_x v - \nabla_x p + T e_{x_3} , \\ \partial_t T + v \cdot \nabla_x T &= \Delta_x T + R v_{x_3} , \\ \nabla_x v &= 0 , \end{aligned}$$

where p is the pressure field, P is the Prandtl number, R is the Rayleigh number (proportional to the temperature difference of top and bottom), v_{x_3} is the vertical component of the velocity field and e_{x_3} is the unit vector in the vertical direction. We use the notation $x \in \mathbf{R}^3$ and assume the components of x are x_1, x_2, x_3. One should also add boundary conditions on the sides.

Since we are dealing with a cubic container we can decompose the

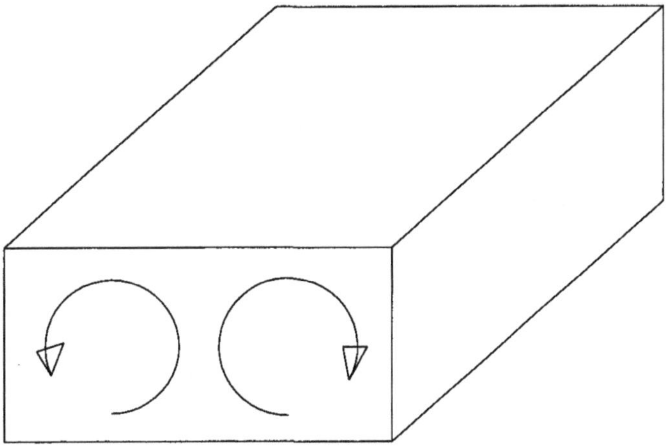

Fig. 1: The experimental setup for the Rayleigh-Bénard experiment.

temperature and velocity fields in Fourier series in the space variables. Taking the Fourier transform of the Boussinesq system we obtain an infinite system of coupled differential equations for the time evolution of the Fourier components. This system is a priori very complicated but we explain now why it is at least formally of *finite dimensional nature*. Consider the equation for the time evolution of a Fourier component of the temperature field. We consider periodic boundary conditions to make the argument simpler. By Fourier components, we mean the decomposition

$$T(x,t) = \sum_{m,n,p\in\mathbf{Z}} T_{m,n,p}(t)\, e^{2\pi i(mx_1+nx_2+px_3)/l} ,$$

where the x_i are the coordinates in $\mathbf{R}^3$. The time evolution equations are then of the form

$$\partial_t T_{m,n,p} = -\left(m^2/l^2 + n^2/l^2 + p^2/l^2\right) T_{m,n,p} + r_{m,n,p}(v,T) ,$$

where $r_{m,n,p}(v,T)$ is some complicated nonlinear expression. For large values of $m^2 + n^2 + p^2$ the first term, which is dissipative, dominates the nonlinear part. In other words, the Fourier modes of T with large indices are strongly damped and their dynamics is essentially driven by the Fourier modes of T with small indices.

To illustrate this point more precisely consider the much simpler differential equation for the unknown function $x : \mathbf{R} \to \mathbf{R}$ given by

$$dx(t)/dt = -Ax(t) + f(t) , \tag{3.1}$$

where $f : \mathbf{R} \to \mathbf{R}$ is some given bounded function of time and A is a large positive number. The solution is

$$x(t) = e^{-At}x(0) + \int_0^t ds\, e^{-A(t-s)} f(s) .$$

If A is large (relative to the time scale on which f changes), we observe that the memory of the initial condition $x(0)$ is rapidly lost. Moreover, the value of $x(t)$ depends only on the values of the function f near t. Therefore, we expect that $x(t)$ tries "to follow" $f(t)$,

$$x(t) \approx f(t)/A .$$

Assume now that the equation (3.1) is replaced by

$$dx(t)/dt = -Ax(t) + F(x(t), t) , \tag{3.2}$$

with $F(x(t), t) = f(t) + x(t)G(x(t), t)$. If A is very large, then it is plausible that the term $x(t)G(x(t), t)$ can be approximated by $f(t)/A \cdot G(f(t)/A, t)$, so that, in some sense, the nonlinearities (in x) can be neglected.

This suggests that if the dissipative part is very positive (one says that A is an elliptic operator), only a few Fourier modes with small indices suffice to describe the time evolution of the complete system, and moreover there should be a closed finite system of differential equations describing their time evolution. This is sometimes called the slaving principle. The mathematically rigorous statement will be given later, it is called the invariant manifold theorem. In the physical literature the irrelevant modes are often called slave modes for obvious reasons.

While the above argument is basically correct, its validity is less useful when the size l of the container is large and the number of relevant modes increases. (For example, for the Laplacian operator on a d dimensional torus, it is easy to verify that the distance between the eigenvalues is of the order of the square of the inverse of the size of the torus. The spectrum of eigenvalues is becoming very dense.) The above reduced system is then as difficult to analyze as the original partial differential equation.

One could therefore be inclined to reject the study of infinite systems as unnecessarily complicated. There are, however, some advantages, because the space translation group acts more naturally on infinite systems than on finite ones, and this is the reason for the increasing interest in the study of such systems.

4. Evolution Equations for Infinite Systems, Constraints

Having argued in the preceding section that infinite systems are of some interest for theoretical studies, we are now confronted with a problem which commonly is absent for systems with finite dimensional phase space: *The existence of solutions for all times*. Later on, we shall only consider equations where this problem does not really arise, but in this section we mention some of the considerations in connection with this problem. The section can thus be skipped without consequences for the further discussion.

Most often, physics is done in Hilbert spaces, but, for the study of dynamical systems, Banach spaces seem more adequate. It would, of course, have been easier (and more familiar to physicists) to use Hilbert spaces instead of Banach spaces. But Hilbert spaces usually involve some decay (or growth) at infinity and we do not want to make such hypotheses. Thus, typical spaces are spaces of functions with bounded derivatives such as $\mathbf{X}_p$, where p a positive integer (or 0) and which are given by

$$\mathbf{X}_p = \{u : u, \partial_x u, \dots, \partial_x^p u \in \mathrm{L}^\infty\} .$$

Consider now the equation

$$\partial_t u = \mathcal{N}(u) . \tag{4.1}$$

In the discussion below, we assume that $\mathcal{N}$ is a real nonlinear differential operator (or pseudo-differential operator) which is quasilinear, and elliptic. (These terms will be explained later.) We also require that $\mathcal{N}$ can be differentiated as often as needed.

We shall assume that the Cauchy problem for the above time evolution is well posed. By this we mean that for a given initial condition

$u(x, 0)$ belonging to one of the above spaces $\mathbf{X}_p$, the solution $u(\cdot, t)$ of (4.1) defines a nonlinear time evolution semigroup S_t which is defined for all $t > 0$. That is,

$$u(x, t) = S_t(u(x, 0))$$

also belongs to $\mathbf{X}_p$. We shall assume that S_t is regular, namely differentiable in the initial condition and also in time; see also [GS].

It may happen that the evolution semigroup is regularizing. By this, one means that the solution is more regular than the initial condition. This is usually a consequence of the domination of the nonlinearity in $\mathcal{N}$ by the highest order elliptic linear part. (In the example of equation (3.2) the operator A is the "elliptic" linear part and F is the nonlinearity.) Proving such a statement can be a mathematically difficult problem. In finite space domains one usually uses Sobolev inequalities which give a (nonlinear) estimate of some L^p norm of a function in terms of its L^2 norm and the L^2 norm of its gradient. This method is, however, not always successful. For example, the estimate is saturated for the two dimensional Navier-Stokes equation and does not work for the three (or higher) dimensional case. Up to now for the three dimensional Navier-Stokes equation in a finite volume one only knows global existence and regularity for small Reynolds numbers and initial conditions. For larger Reynolds numbers one can only prove uniqueness and regularity of the initial value problem for a finite time interval or for small initial conditions. There are some problems like the two dimensional nonlinear Schrödinger equation where one is even able to prove that some initial conditions give solutions which blow up in finite time. Of course, this is not a very physical situation but it indicates that some stabilizing mechanism has been neglected in the model.

We also mention that, even though an evolution equation may not be well behaved (or not known to be so), one may find interesting and physically acceptable solutions that are worthy of consideration; see [M4].

The question is even more difficult to settle for systems in unbounded space domains. One can sometimes localize spatially the problem and use the above techniques. We shall not expand on these rather technical questions and assume that the dynamics is regular and well defined.

One meets very often *equations with constraints* which restrict the state function u. These are, in general, associated to conservation laws. A well-known example is the incompressibility condition in fluid dynamics (that is, the equation (1.1)). We shall model the constraints by a (nonlinear) differential (or pseudo-differential) operator $\mathcal{C}$. The constraints are given by the equation

$$\mathcal{C}(u) = 0 .$$

We assume that $\mathcal{C}(\cdot)$ has the same properties as $\mathcal{N}$. In particular, it can be differentiated as many times as we need. We also assume that the constraint $\mathcal{C}$ is maintained by the time evolution. More precisely, if the initial condition u belongs to the domain of $\mathcal{C}$, and satisfies $\mathcal{C}(u) = 0$, then for every positive time t we must have

$$\mathcal{C}\left(S_t(u)\right) = 0 .$$

In the example of the incompressible Navier-Stokes equation, the pressure cancels the divergence of the nonlinearity, and one gets a diffusion equation for the evolution of the divergence of the speed. Therefore if this divergence is zero initially it will remain zero at positive times.

In some accordingly rather vague sense, the above situation defines a dynamical system (with constraints). In the sequel, we shall follow the ideas and organization of the standard theory and *omit* the question of constraints in the study of problems in infinite domains.

5. Model Equations

We have seen that there are, roughly speaking, two kinds of problems, finite dimensional and infinite dimensional ones. Each has advantages and disadvantages. In both cases, one would like to classify the various possibilities for the large time evolution, the goal being to simplify the analysis by collecting in the same class some apparently unrelated problems. We recall that a similar program is indeed successful in the study of second order phase transitions in Statistical Mechanics. As we shall see below this classification program is essentially completed for the study of systems of coupled differential equations using the notion of normal forms. The main idea is in a vague sense to put together all the time evolutions with equivalent dynamics and to describe each such class by one system which is hopefully simpler than the original one. For the case of systems in infinite space, this program is far from its end, but some of the important simple systems have already been recognized. The most frequent example is the so-called Newell-Whitehead equation, which is also called the Ginzburg-Landau equation, or amplitude equation. It is given by

$$\partial_t u(x,t) = \partial_x^2 u(x,t) + u(x,t) - u(x,t)|u(x,t)|^2 , \qquad (CA)$$

where u is a complex field. There is also a real version without the absolute value which is often called the nonlinear heat equation,

$$\partial_t u(x,t) = \partial_x^2 u(x,t) + u(x,t) - u^3(x,t) . \qquad (RA)$$

Another important example (which is not a normal form) is the Swift-Hohenberg equation

$$\partial_t u(x,t) = \alpha u(x,t) - (1+\partial_x^2)^2 u(x,t) - u^3(x,t) \ , \qquad (SH)$$

where u is a real field and α is a parameter. A third example is the Kuramoto-Sivashinsky equation

$$\partial_t u = -u^{(4)} + \lambda u^{(2)} + uu' \ .$$

Other examples are the nonlinear Schrödinger equation, and the Zakharov equations of plasma physics. We shall explain later how these equations are derived from the physical equations mentioned above and the sense in which they describe the interesting part of the dynamics.

Remark. Throughout this book, the abbreviations RA, CA, and SH will denote the real amplitude equation, the complex amplitude equation and the Swift-Hohenberg equation, respectively.

We will attempt to derive the "genericity" of the above equations in Section 23.

6. Fronts

A very common problem for systems which are not in equilibrium can be formulated as follows. Suppose a substance can exist in two phases. Assume an initial condition has been prepared such that the two phases coexist in different regions of space with a common boundary (interface). What is the future evolution for the system? This problem arises for example in solidification (dendrites, spinodal decompositions and droplets, etc.), but also in hydrodynamics (spatial intermittency), chemistry, biology etc. If one considers a sharp interface one is led to a moving boundary problem. But very often the boundary is not sharp (consider for example temperature fields, concentration fields etc.). The physical process driving the time evolution of the system is usually described by dissipative (diffusive) nonlinear partial differential equations (the Navier-Stokes equation is a typical example of nonlinearity). These equations are of parabolic type and, as explained earlier, we shall assume that they define a well behaved semigroup of time evolutions on a space of regular and bounded functions. Since we are working with infinite spatial domains this result may not be so easy to prove.

Up to now the problem has been mathematically analyzed mostly in one dimension (except for the recent results on dendrites and the Saffman-Taylor problem). In space dimension larger than one, a major open problem is to understand the instabilities of the interface which can end up being a fractal curve. In view of this difficulty, we will only discuss one dimensional problems.

As explained above, we shall investigate the question for time evolution of an initial condition describing the coexistence of two different phases. We shall adopt here a rather conservative definition of phases. Namely, we shall call a phase any stationary solution of the evolution equation. For the RA equation the stationary solutions $u = 0$ and $u = \pm 1$ are phases. There is a noticeable difference between them. The solution $u = 0$ is linearly unstable, while each solution $u = \pm 1$ is linearly stable. Therefore, we expect that an initial condition $u(x, 0)$ interpolating between 1 and 0 ($u(-\infty, 0) = 1$ and $u(+\infty, 0) = 0$) for example will evolve toward the constant solution 1. In other words, the domain occupied by the more stable phase will grow at the expense of the domain occupied by the least stable one.

We will explain this in detail in Chapter VII. An even more interesting situation occurs when the stationary solution is spatially inhomogeneous. Then, as the front passes by, it will leave a pattern in the laboratory frame. This provides therefore an example of spontaneous pattern formation.

CHAPTER II
SMALL SOLUTIONS

We are, of course, interested in nontrivial solutions of the various equations mentioned in Chapter I. We will obtain various such solutions in the different stages of bifurcation theory developed in this book. Before we start with this program, we wish to prepare some material which will be needed essentially every time we talk about "small" solutions, or about the time evolution of small initial conditions. Intuitively, for small solutions, homogeneous nonlinearities (such as u^3 in the real amplitude equation) can be dropped and are "irrelevant" for the understanding of what is going on. The theory of **invariant manifolds** makes these notions precise in an efficient way.

7. Invariant Manifolds

In this section, we consider invariant manifolds at a hyperbolic fixed point of a vector field X.

Notation for derivatives. In this book, we use the following notation for derivatives: either the symbol ∂_x, which means the derivative with respect to the variable x, or D. If f is a function from, say, $\mathbf{R}^d$ to $\mathbf{R}^n$, then $\partial_x f$ means the $d \times n$ matrix of partial derivatives of f. This matrix is sometimes also denoted $\mathrm{D}f$. If f is a function of two variables, say,

$$f : (x, y) \mapsto f(x, y) ,$$

then we denote the (matrix of) partial derivative(s) by $\partial_x = \mathrm{D}_1 = \mathrm{D}_x$, $\partial_y = \mathrm{D}_2 = \mathrm{D}_y$, and similarly if there are more variables. If f is a function of λ then the notation $\partial_\lambda f(0)$ (respectively $\partial_\lambda f_0$ when the argument appears as an index) is a shorthand notation for $(\partial_\lambda f)(0)$.

Let $X : \mathbf{R}^d \to \mathbf{R}^d$ be a real vector field, which is as often differentiable as we need (the precise conditions will be spelled out later). Assume $X(0) = 0$ (that is, 0 is a fixed point for the flow) and assume that the eigenvalues of the matrix $\mathrm{D}X(0)$ (at 0) have all different real part. To be specific, we consider the following example for $d = 2$. Denote by

x_1 and x_2 the two components of an element x of $\mathbf{R}^2$. Then the vector field $X(x)$ has the two components

$$\begin{aligned} X_1(x) &= x_2 , \\ X_2(x) &= -cx_2 + x_1^3 - x_1 , \end{aligned} \tag{7.1}$$

where $c > 2$. Clearly, $X(0) = 0$ and the tangent matrix at 0 is

$$\mathrm{D}X(0) = \begin{pmatrix} 0 & 1 \\ -1 & -c \end{pmatrix} .$$

The eigenvalues are

$$\lambda_\pm = \frac{-c \pm \sqrt{c^2 - 4}}{2} ,$$

and the corresponding eigenspaces are denoted by $E_\pm$.

Consider now the differential equation

$$\partial_t x(t) = X(x(t)) . \tag{7.2}$$

This vector field is the ordinary differential equation associated with solutions $u(y, t) = v(y - ct)$ of the real amplitude equation

$$\partial_t u = \partial_x^2 u + u - u^3 .$$

Then v satisfies

$$v'' = -cv' + v^3 - v ,$$

and setting $v(\tau) = x_1(\tau)$ and $v'(\tau) = x_2(\tau)$ we obtain the equation (7.2) for the vector field (7.1), with $t = \tau$.

The *invariant manifold theorem* says that there are two curves, $W_\pm$, tangent to $E_\pm$ at 0, which are solution curves to the Equation (7.2). The invariance means that if $x(t_0) \in W_\pm$ then $x(t)$ is in $W_\pm$ for all t. In the case of W_-, the definition implies that for asymptotic times, $t \to +\infty$, the motion is asymptotic to a motion of the linearized equation

$$\partial_t x(t) = (\mathrm{D}X)(0) \cdot x(t) ,$$

where $\cdot$ denotes the matrix action on the vector x. Similarly, the same statement holds for W_+, when $t \to -\infty$.

If the eigenvalues of $\mathrm{D}X(0)$ have opposite real part then *every* initial point close to 0 gets pulled into W_+ (or E_+) and leaves a small neighborhood of the origin except for those points lying on the invariant manifold W_- (assuming W_- is associated with the eigenvalue whose real part is negative). Therefore, W_- is called the stable manifold of X (at 0) and W_+ is called the unstable manifold. This is illustrated in Fig. 2.

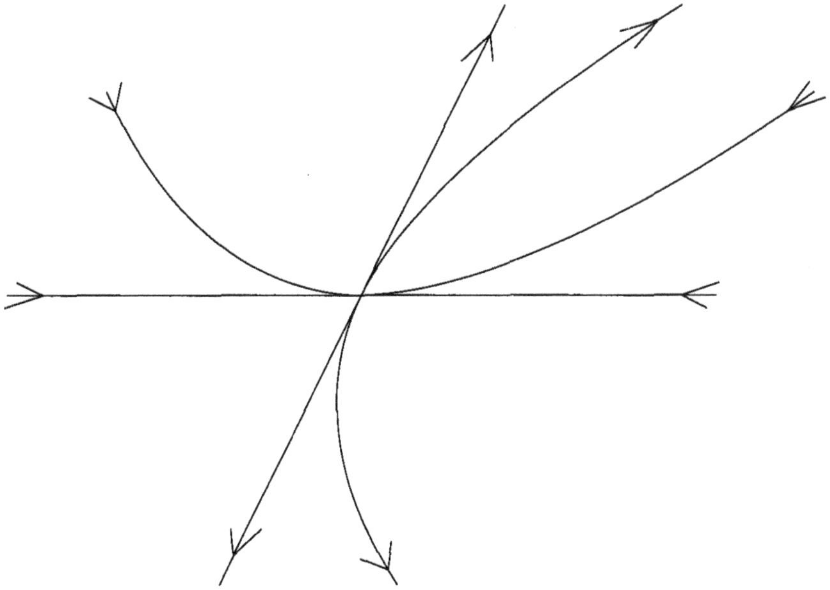

Fig. 2: A hyperbolic fixed point for a flow with the stable manifold tangent to the stable eigenspace.

Remark. The invariant manifolds are *unique* in the hyperbolic case, and, in general, nonunique in the other cases. The reason for the nonuniqueness is the following, cf. Fig. 3. Every initial point will be pulled into a curve which is tangent to the *weaker* eigenvalue, except those lying on the unique invariant manifold associated with the stronger eigenvalue.

We now give a very general formulation of the above ideas.

Theorem 7.1. *Let Y be a C^k vector field in $\mathbf{R}^d$ with a stationary solution at the origin. Let Σ be the spectrum (finite set of eigenvalues) of the real matrix $\mathrm{D}Y(0)$. Assume that Σ is the disjoint union of two nonempty sets Σ_1 and Σ_2 and assume there are a real number $\sigma \leq 0$ and an integer $1 \leq r \leq k$ such that*

$$\operatorname{Re}\Sigma_2 < r\sigma \leq \sigma < \operatorname{Re}\Sigma_1 \ . \tag{7.3}$$

Then there is an invariant C^r manifold W_1 through 0 of dimension equal to the cardinality of Σ_1. This manifold is tangent at the origin to the spectral subspace of $\mathrm{D}Y(0)$ corresponding to Σ_1 (and similarly for Σ_2). Moreover, the invariant manifold W_1 is locally attracting

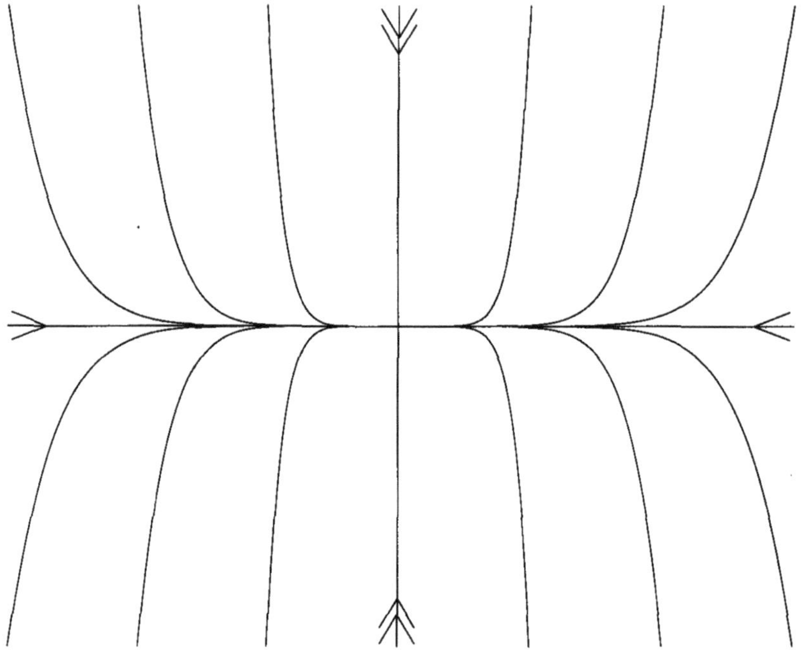

Fig. 3: The flow for a vector field with a fixed point whose two eigenvalues attract. Most initial points are on curves tangent to the weaker eigendirection.

near 0, namely the transverse dynamics is topologically equivalent to the linear equation

$$dz/dt = -z ,$$

where z is a transverse coordinate to W_1 (of dimension the cardinality of Σ_2).

Analogous statements hold when $\sigma \geq 0$, and the condition (7.3) has to be replaced by

$$\mathrm{Re}\,\Sigma_2 < \sigma \leq r\sigma < \mathrm{Re}\,\Sigma_1 \ .$$

Note that when $\sigma = 0$, we can always choose $r = k$.

The exact definition of topological equivalence will be given in Section 12.

This theorem is also true for infinite dimensional systems in Banach spaces. It implies that near the origin the dynamics is pulled in to the invariant manifold W_1 and is then very similar to the evolution on the invariant manifold W_1 itself. The interesting case of this theorem occurs when the dimension of W_1 is much smaller than d. In particular, when d is infinite, the dimension of W_1 can still be finite.

The proof follows from the implicit function theorem, we refer to [HPS] for the details. A related result will be proven in Section 9.

If we have a family of vector fields $X(\lambda, \cdot)$ which depends smoothly on some parameter λ (which can be a vector), we can try to apply the above theorem for each value of the parameter. There is, however, an elementary extension of the above theorem which allows to treat simultaneously an open set of parameter values. If the vector field X is acting in $\mathbf{R}^d$ and the parameter λ belongs to $\mathbf{R}^\nu$, we can consider the vector field Y of $\mathbf{R}^{\nu+d} = \mathbf{R}^\nu \otimes \mathbf{R}^d$ given by

$$Y(\lambda, x) = (0, X(\lambda, x)) .$$

Assume that the origin is a stationary solution of $X(0, \cdot)$. We can now apply Theorem 7.1 to the vector field Y at the stationary solution $(0, 0)$ and we obtain the following result. Denote by Σ_1 the eigenvalues of $D_2X(0, 0)$ lying on the imaginary axis. Let P_1 be the corresponding spectral projection, and let $P_2 = 1 - P_1$, and denote by E_1 and E_2 the corresponding spectral subspaces.

Theorem 7.2. Center Manifold Theorem. *For every λ near 0 there exists a manifold $W_\lambda^0 \subset \mathbf{R}^d$ with the following properties:*

1) *W_0^0 is tangent to E_1 at $0 \in \mathbf{R}^d$.*
2) *The manifold W_λ^0 is locally invariant under the flow of the vector field $X(\lambda, \cdot)$.*
3) *There is a neighborhood U of 0 in $\mathbf{R}^d$, which is independent of λ such that every point whose orbit stays forever in U is attracted to W_λ^0.*
4) *If X depends smoothly on λ and x, then W_λ^0 is a smooth family of manifolds.*

For a proof, see for example Ruelle [R2].

8. Perturbation Theory

We shall now explain how to compute the formal power series for the invariant manifold. We shall assume that Y is a $\mathcal{C}^\infty$ vector field on $\mathbf{R}^d$ with zero as a singular point (in other words $Y(0) = 0$). Let Σ_1 (resp. Σ_2) be the set of eigenvalues of $\mathrm{D}Y(0)$ contained in the closed right half-plane (resp. open left half-plane). We shall denote by P_1 and $P_2 = 1 - P_1$ the corresponding spectral projections, by A_1 and A_2 the restrictions of $\mathrm{D}Y(0)$ to the two spectral subspaces E_1 and E_2. We shall from now on identify $\mathbf{R}^d$ with the direct sum of these two subspaces. We shall look for the invariant manifold as a graph of a function $y \mapsto z(y)$ from E_1 to E_2 which is tangent to E_1 at the origin. Considering the time evolution of a vector $(y, z(y))$ on the invariant manifold and projecting on the two spectral subspaces we get the following equations

$$\begin{aligned} dy/dt &= P_1 Y(y, z(y)) , \\ \mathrm{D}z(y) dy/dt &= P_2 Y(y, z(y)) . \end{aligned}$$

Using the first equation we can eliminate dy/dt in the second equation and we get a closed equation for z:

$$\mathrm{D}z(y) P_1 Y(y, z(y)) = P_2 Y(y, z(y)) . \tag{8.1}$$

If we have solved this equation for z we can use this information in the first differential equation and we get a closed time evolution for the variable y.

We next solve the equation for the unknown function z in formal power series. To simplify the notation, we assume that E_1 is the subspace

$$E_1 = \{ y \in \mathbf{R}^d : y_{s+1} = 0, \ldots, y_d = 0 \} ,$$

and

$$E_2 = \{ y \in \mathbf{R}^d : y_1 = 0, \ldots, y_s = 0 \} .$$

Then the formal power series for z is of the form

$$z(y) = \sum_p z_p(y)$$

with

$$z_p(y) = \sum_{\substack{i_1, \ldots, i_s \geq 0 \\ i_1 + \cdots + i_s = p}} z_p^{(i_1, \ldots, i_s)} y_1^{i_1} \cdots y_s^{i_s} ,$$

where the coefficients $z_p^{(i_1,\dots,i_s)}$ are vectors in E_2. We want to find z such that it is a formal solution of the equation (8.1). We shall find a similar expansion in Section 12. If we further assume for simplicity that $DY(0)$ is diagonal in the basis used above, we can write an equation for the components $z_{p,j}$ of z_p. We first observe that $z_0 = 0$ because the invariant manifold passes through the origin. We have also $z_1 = 0$ because the invariant manifold is tangent to E_1 at the origin. Therefore, z is quadratic in y and the first nontrivial equation starts with $p = 2$. It is easy to verify that the equations can be solved recursively. For simplicity we assume that $DY(0)$ has only simple eigenvalues ρ_j ($j = 1, \dots, d$), and is diagonal in the natural basis on $\mathbf{R}^d$. The pth equation written in coordinates is then

$$\left(\rho_j - \sum_{\ell=1}^{s} i_\ell \rho_\ell\right) z_{p,j}^{(i_1,\dots,i_s)} = \mathcal{F}_{p,j} , \qquad \text{for } j = s+1, \dots, d ,$$

where $\mathcal{F}_p$ is a function of the derivatives of Y at 0 up to order p and of the already computed Taylor coefficients $z_2, \dots, z_{p-1}$. From our previous assumptions, if the eigenvalue ρ belongs to Σ_1 we have $\operatorname{Re}\rho \geq 0$, and there is a positive number σ such that if ρ belongs to Σ_2 we have $\operatorname{Re}\rho < -\sigma$. Therefore, in the above equation the coefficient of $z_{p,j}^{(i_1,\dots,i_s)}$ has an absolute value bounded below by σ and we can solve for $z_{p,j}^{(i_1,\dots,i_s)}$. We will see later that there are problems in which these coefficients can vanish or become very small. This is called a resonance or a small divisor problem.

9. Stable Manifolds (Unbounded Case)

In Chapter VII, we will need a generalization of the above results for the case when the linearized problem leads to an unbounded operator in a Banach space. This section will not be used until Chapter VII and can thus be skipped at first reading.

The existence theory for stable and unstable manifolds in an infinite-dimensional setting with a *bounded* linear part is a well-studied subject, cf. [HPS], [La]. We do not repeat it here, but we give a proof for the more complicated situation of an *unbounded* linear part. The more simple situations follow readily, but more elegant proofs are available in those cases.

When the linear operator A is unbounded, *the time evolution may not be defined for all positive* t, unless one restricts attention to a semigroup picture. This is what will be done now.

We consider – to begin with, on a purely formal level – an equation of the form

$$\frac{dX}{dt} = AX + F(X), \tag{9.1}$$

with X in some separable Banach space $\mathbf{B}$, A unbounded, but with spectrum away from the imaginary axis. The map F maps a neighborhood of 0 in $\mathbf{B}$ to $\mathbf{B}$ and is a twice Fréchet differentiable function satisfying $F(0) = 0$, $F'(0) = 0$. Intuitively, this means that F is quadratic in X. The Eq. (9.1) has $X = 0$ as a solution, and we are interested in the stable manifold tending to this solution.

Assumptions. We describe in detail the assumptions on A and F.

A1. We assume that A has spectrum bounded away from the imaginary axis. We do not assume that A is bounded.

A2. Denote by P^s and P^u the spectral projections onto the subspaces E^s and E^u corresponding to the part of the spectrum with negative and positive real part, respectively. We assume that these projections exist and are bounded operators (cf. for example [K]). We define the operators $A^s = P^s A$ and $A^u = P^u A$, and we assume that they have dense domain in E^s and E^u and that their spectra are bounded away from the imaginary axis. We further assume that A^s and $-A^u$ are the generators of equicontinuous semigroups of Class C_0 in the sense of Yosida [Y].

A3. We denote by $\|\cdot\|_s$ and $\|\cdot\|_u$ the induced norms on E^s and E^u. We make the following continuity and differentiability assumptions on F: There is a constant $D < \infty$ such that

$$\begin{aligned} \|F(x,y)\| &\le D \max(\|x\|_s, \|y\|_u)^2 , \\ \|DF(x,y)(x',y')\| &\le D \max(\|x\|_s, \|y\|_u) \max(\|x'\|_s, \|y'\|_u) , \end{aligned} \tag{9.2}$$

on a neighborhood of $(0,0)$.

Remark. The conditions in A2 can also be formulated in terms of the resolvents. For example, it is sufficient that there exists a constant such that $\|(A^s - \xi)^{-k}\|_s <$ const. ξ^{-k}, for $k = 1,2,\ldots$, when the real quantity $\xi > 0$ is not in the spectrum, cf. [K].

By changing (if necessary) the norm to an equivalent new norm $\|\cdot\|_s'$ defined by

$$\|x\|_s' = \sup_{t\ge 0} \|e^{(A^s - \lambda^s)t} x\|$$

(and analogously for A^u), we may assume that the semigroups satisfy the bounds

$$\begin{aligned} \sup_{t\ge 0} e^{\lambda^s t} \|e^{A^s t}\|_s &\le 1 , \\ \sup_{t\ge 0} e^{\lambda^u t} \|e^{-A^u t}\|_u &\le 1 , \end{aligned} \tag{9.3}$$

with $\lambda^s > 0$, $\lambda^u > 0$. If A is selfadjoint, then λ^s and λ^u can be chosen as the distances of the spectrum from the imaginary axis. By A2, we have that P^s and P^u are bounded, and we choose D such that $\|P^s\|_s < D$ and $\|P^u\|_u < D$.

Although the operators A^s and A^u may be unbounded, the construction of W^s is possible because it can be formulated in terms of the semigroups $e^{A^s\tau}$, $e^{-A^u\tau}$, $\tau \ge 0$, and these semigroups *are* defined. To formulate the problem in terms of these semigroups, we consider the evolution equations on the stable manifold itself.

To do so, one considers W^s as the graph of a function $\Phi : E^s \to E^u$. We shall prove the existence of Φ on a small ball $\mathbf{B}_r^s$ in E^s (of radius r, centered at 0). Assume momentarily that Φ is already known. Then we can express the flow on W^s (where it will be seen to be defined for all $t \ge 0$) as follows: Denote by ψ_t^Φ the solution (for fixed Φ) of

$$\frac{d}{dt}\psi_t^\Phi = A^s \psi_t^\Phi + G_\Phi^s \circ \psi_t^\Phi , \tag{9.4}$$

where $\psi_t^\Phi : E^s \to E^s$, and

$$G_\Phi^s(y) = P^s F(y, \Phi(y)),\ y \in E^s .$$

If Φ is the graph of W^s, then ψ_t^Φ is the flow in W^s, projected onto E^s. But we define ψ_t^Φ by (9.4) even if Φ is arbitrary. The solution of (9.4) is formally given by

$$\psi_t^\Phi = e^{A^s t} + \int_0^t d\tau\, e^{A^s(t-\tau)} G_\Phi^s \circ \psi_\tau^\Phi . \tag{9.5}$$

We now study the condition that Φ describes W^s. In order to stay on W^s, the flow ϕ_t on $E^u \oplus E^s$ must be of the form

$$\phi_t \begin{pmatrix} x \\ \Phi(x) \end{pmatrix} = \begin{pmatrix} \psi_t^\Phi(x) \\ \Phi(\psi_t^\Phi(x)) \end{pmatrix} ,$$

which, upon differentiation with respect to t and restriction to the second component, leads to the equation (on E^s)

$$D\Phi_x \cdot \frac{d}{dt}\psi_t^\Phi \bigg|_{t=0} = A^u \Phi + G_\Phi^u , \tag{9.6}$$

where $G_\Phi^u(x) = P^u F \begin{pmatrix} x \\ \Phi(x) \end{pmatrix}$.

The equation (9.6) follows from

$$\Phi = - \int_0^\infty d\tau\, e^{-A^u \tau} G_\Phi^u \circ \psi_\tau^\Phi . \tag{9.7}$$

We shall solve the existence problem for the stable manifold by considering the system (9.5), (9.7) in the (coupled) unknowns ψ_τ^Φ and Φ. We reformulate the problem as a fixed point problem and solve it in a suitable function space, inspired by methods from [HPS].

Given Φ, we solve (9.5), for ψ_τ^Φ and then we substitute that solution into the r.h.s. of (9.7). The resulting quantity will be called $C(\Phi)$ and the problem will be solved if we can find a Φ for which $C(\Phi) = \Phi$. We shall show that the operator C is a contraction of a sufficiently small ball of functions, so that the equation $C(\Phi) = \Phi$ has a (locally unique) solution by the contraction mapping principle.

We shall fix below a small $r > 0$, and a $\sigma > 0$. We write $\mathbf{B}_r^s$ for the ball of radius r in E^s centered at 0, that is,

$$\mathbf{B}_r^s = \{x \in E^s : \|x\|_s \le r\} .$$

We write $\mathbf{B}_r^u$ for the ball of radius r in E^u. We define next the set in which Φ will live:

$$\mathbf{A}_{r,\sigma} = \{\Phi : \mathbf{B}_r^s \to E^u,\ \Phi(0) = 0\ ,\\ \|\Phi(x) - \Phi(y)\|_u \le \sigma\|x - y\|_s, \text{ for } x, y \in \mathbf{B}_r^s\}\ .$$

We define, for Φ, $\Psi \in \mathbf{A}_{r,\sigma}$, the distance (see [HPS])

$$d_{\mathbf{A}_r}(\Phi, \Psi) = \sup_{\substack{x\in\mathbf{B}_r^s\\ x\neq 0}} \frac{\|\Phi(x) - \Psi(x)\|_u}{\|x\|_s}\ .$$

Equipped with this distance, $\mathbf{A}_{r,\sigma}$ is a complete metric space.

We next define a set $\mathbf{B}_{r,\sigma}$ in which the flow ψ^Φ lives:

$$\mathbf{B}_{r,\sigma} = \{\psi : \mathbf{R}^+ \times \mathbf{B}_r^s \to \mathbf{B}_r^s\,,\ \psi_0(x) = x\ ,\ \psi_t(0) = 0 \text{ for all } t > 0\ ,\\ \sup_{t\ge 0} \|\psi_t(x) - \psi_t(y)\|_s \le \sigma\|x - y\|_s\}\ .$$

We equip the set $\mathbf{B}_{r,\sigma}$ with a distance

$$d_{\mathbf{B}_r}(\psi, \psi') = \sup_{t\ge 0} \sup_{\substack{x\in\mathbf{B}_r^s\\ x\neq 0}} \frac{\|\psi_t(x) - \psi'_t(x)\|_s}{\|x\|_s}\ .$$

It is easy to verify that $\mathbf{B}_{r,\sigma}$ is a complete metric space for this distance.

Proposition 9.1. *For every $\sigma > 1$ one can choose a sufficiently small $r > 0$ so that the system (9.5), (9.7) has a unique solution Φ, ψ_r^Φ in the spaces $\mathbf{A}_{r,\sigma}$ and $\mathbf{B}_{r,\sigma}$.*

Theorem 9.2. *Under the Assumptions A1 – A3, the dynamical system (9.1) has a stable manifold W^s which is the graph of the map $\Phi : E^s \to E^u$.*

Proof. This is an immediate consequence of Proposition 9.1.

Proof of Proposition 9.1. We first fix an arbitrary $\Phi \in \mathbf{A}_{r,\sigma}$. We start by showing the existence and studying the properties of a function ψ^Φ, which solves (9.5).

Assume that $\psi \in \mathbf{B}_{r,\sigma}$, and $\Phi \in \mathbf{A}_{r,\sigma}$ and denote by $H_\Phi(\psi)$ the r.h.s. of (9.5), that is,

$$H_\Phi(\psi)_t = e^{A^s t} + \int_0^t d\tau\, e^{A^s(t-\tau)} G^s_\Phi \circ \psi_\tau \ . \tag{9.8}$$

From the definitions, we have

$$\begin{aligned}\|\psi_\tau(x) - \psi_\tau(y)\|_s &\le \sigma\|x-y\|_s \\ \|\Phi(\psi_\tau(x)) - \Phi(\psi_\tau(y))\|_u &\le \sigma\|\psi_\tau(x) - \psi_\tau(y)\|_s \ ,\end{aligned}$$

and, by the normalizations,

$$\|\Phi(\psi_\tau(x))\|_u \le \sigma\|\psi_\tau(x)\|_s \le \sigma^2\|x\|_s \ .$$

This implies

$$\begin{aligned}\max(\|\psi_\tau(x)\|_s, \|\Phi(\psi_\tau(x))\|_u) &\le \max(\|\psi_\tau(x)\|_s, \sigma\|\psi_\tau(x)\|_s) \\ &\le (\sigma+\sigma^2) r \ .\end{aligned}$$

Therefore, we have

$$\begin{aligned}&\|H_\Phi(\psi)_t(x) - H_\Phi(\psi)_t(y)\|_s \\ &\quad\le \left(e^{-\lambda^s t} + \int_0^t d\tau\, e^{-\lambda^s(t-\tau)} D^2 r(\sigma+\sigma^2)^2 \right) \|x-y\|_s \ .\end{aligned} \tag{9.9}$$

Setting $y = 0$ this implies that the function $H_\Phi \circ \psi_\tau$ maps the ball $\|x\|_s < r$ in E_s to itself, provided

$$rD^2(\sigma+\sigma^2)^2 < \lambda^s \ .$$

Note also that by (9.9) we see that $H_\Phi(\psi) \in \mathbf{B}_{r,\sigma}$ if

$$1 + \frac{rD^2(\sigma+\sigma^2)}{\lambda_s} < \sigma \ .$$

These inequalities are clearly satisfied for sufficiently small r.

We next show that H_Φ is a contraction. We have

$$\begin{aligned}&\|H_\Phi(\psi)_t(x) - H_\Phi(\psi')_t(x)\|_s \\ &\le \int_0^t d\tau\, e^{-\lambda^s(t-\tau)} D^2(1+\sigma)(\sigma+\sigma^2) r \|\psi_\tau(x) - \psi'_\tau(x)\|_s \\ &\le \sigma r D^2(1+\sigma)^2 \|x\|_s \int_0^t d\tau\, e^{-\lambda^s(t-\tau)}\, d_{\mathbf{B}_r}(\psi,\psi') \\ &\le \sigma r D^2(1+\sigma)^2 \|x\|_s \frac{1}{\lambda^s}\, d_{\mathbf{B}_r}(\psi,\psi') \\ &\le \frac{1}{2}\|x\|_s\, d_{\mathbf{B}_r}(\psi,\psi') \ ,\end{aligned}$$

provided

$$2\sigma r D^2(1+\sigma)^2 < \lambda^s .$$

If we choose r sufficiently small, then all of the above inequalities hold. This implies that H_Φ is a contraction on $\mathbf{B}_{r,\sigma}$ and has therefore a unique fixed point in $\mathbf{B}_{r,\sigma}$. We call ψ^Φ the fixed point one obtains in this fashion.

We next study the map C defined by the r.h.s. of (9.7). We first show that C maps $\mathbf{A}_{r,\sigma}$ into itself. If $\Phi \in \mathbf{A}_{r,\sigma}$, then we have, as before,

$$\|\Phi \circ \psi_\tau^\Phi(x) - \Phi \circ \psi_\tau^\Phi(y)\|_u \le \sigma^2 \|x-y\|_s ,$$

and therefore,

$$\|C(\Phi)(x) - C(\Phi)(y)\|_u \le \frac{rD^2(\sigma+\sigma^2)^2}{\lambda^u}\|x-y\|_s \le \sigma\|x-y\|_s ,$$

for sufficiently small r. In other words, C maps $\mathbf{A}_{r,\sigma}$ to itself. We next show that C is a contraction. Note that

$$G_\Phi^u \circ \psi_\tau^\Phi - G_\Psi^u \circ \psi_\tau^\Psi = (G_\Phi^u \circ \psi_\tau^\Phi - G_\Phi^u \circ \psi_\tau^\Psi) + (G_\Phi^u \circ \psi_\tau^\Psi - G_\Psi^u \circ \psi_\tau^\Psi) .$$

Using estimates of the same form as above, we find

$$\|C(\Phi)(x) - C(\Psi)(x)\|_u \le \frac{D^2(\sigma+\sigma^2)r}{\lambda^u}$$

$$\times \left((1+\sigma) \sup_{\tau\ge 0} \|\psi_\tau^\Phi(x) - \psi_\tau^\Psi(x)\|_s + \sup_{\tau \ge 0} \|\Phi(\psi_\tau^\Psi(x)) - \Psi(\psi_\tau^\Psi(x))\|_u \right)$$

$$\le \frac{D^2(\sigma+\sigma^2)r}{\lambda^u} \left((1+\sigma) \sup_{\tau\ge 0} \|\psi_\tau^\Phi(x) - \psi_\tau^\Psi(x)\|_s + \|x\|_s d_{\mathbf{A}_r}(\Phi,\Psi) \right) .$$

We use Eq. (9.5) to estimate $\psi_\tau^\Phi(x) - \psi_\tau^\Psi(x)$. We have

$$\|\psi_t^\Phi(x) - \psi_t^\Psi(x)\|_s \le \int_0^t d\tau\, e^{-\lambda^s(t-\tau)} \|G_\Phi^s \circ \psi_\tau^\Phi(x) - G_\Psi^s \circ \psi_\tau^\Psi(x)\|_s .$$

From this we get

$$\sup_{t\ge 0} \|\psi_t^\Phi(x) - \psi_t^\Psi(x)\|_s$$

$$\le \frac{D^2(\sigma+\sigma^2)r}{\lambda^s} \left((1+\sigma) \sup_{\tau\ge 0} \|\psi_\tau^\Phi(x) - \psi_\tau^\Psi(x)\|_s + \|x\|_s d_{\mathbf{A}_r}(\Phi,\Psi) \right) .$$

If $D^2(\sigma+\sigma^2)(1+\sigma)r < \lambda^s/2$, then we get

$$\sup_t \|\psi_\tau^\Phi(x) - \psi_\tau^\Psi(x)\|_s \leq 2\frac{D^2(\sigma+\sigma^2)r}{\lambda^s}\|x\|_s d_{\mathbf{A}_r}(\Phi, \Psi) ,$$

and taking r smaller, if necessary, we see that C is a contraction. Therefore, C has a fixed point, as asserted. This fixed point defines the graph Φ, which is describes the stable manifold of the flow (9.1). The proof of Proposition 9.1 is complete.

Remark. A straightforward extension of the above methods, see [HPS], shows that if F is k times Fréchet differentiable, then so is Φ. Note that the bound on Φ will always contain a factor r, so that the stable manifold is close to E^s near the origin.

We now make one more assumption on the linearized vector field at the origin.

A4 The real part of the spectrum of A is bounded.

Corollary 9.3. *Under the hypotheses A1 – A4 there is an invariant unstable manifold* W^u.

We recall that this is an invariant manifold on which the inverse time evolution is contracting to the origin.

The proof is very easy. The differential equation

$$\frac{dX}{dt} = -AX - F(X)$$

satisfies the assumptions A1–A4. Therefore, we can apply Proposition 9.1 and we obtain a stable manifold. This is, however, equation (9.1) for a reversed time. This stable manifold is therefore the unstable manifold for the initial time evolution. The remarks made on the regularity of the stable manifold are, of course, still valid here.

We mention that assumption A4 is needed only to have a well defined time evolution on the unstable manifold. The above argument still holds for operators whose real part of the spectrum is unbounded, however, for such operators the time evolution for a generic initial condition on the unstable manifold is not defined (it diverges immediately).

Perturbation Theory. We explain how to get a perturbation theory (formal power series) for the invariant manifolds constructed above. First of all, the application of the contraction mapping principle always leads to perturbative formulas. Thus they could be derived easily from the developments in this section. A more specific example has already been given in Section 8.

CHAPTER III
BIFURCATION THEORY

In this chapter, we review some of the standard material on bifurcation theory. The experienced reader may skip most of this material. Many good books on the subject exist, and we include the results mostly to fix terminology and notation. Some reference books are: [Ar1], [BPV], [GH], [IJ], and [R2].

10. The Persistence of Point Attractors

We consider changes of a point attractor when external parameters vary. We shall first enlarge somewhat the discussion by analyzing the changes of stationary solutions of a *vector field* which depends on parameters. A general answer is given by the following theorem.

Theorem 10.1. *Let* $X(\lambda, \cdot)$ *be a* $\mathcal{C}^1$ *(once continuously differentiable) real vector field depending* $\mathcal{C}^1$ *on real parameters* λ *(i.e., a vector valued function* $X(\lambda) : \mathbf{R}^d \to \mathbf{R}^d$*). Assume that* $X(0,0) = 0$*, and that the real matrix*

$$\mathrm{D}_2 X(0,0)$$

is invertible. Then there is a neighborhood $\mathcal{V}$ *of* 0 *in parameter space such that for any* $\lambda \in \mathcal{V}$ *there is a unique stationary solution* $M(\lambda)$ *near* 0*. The map*

$$\lambda \mapsto M(\lambda)$$

is $\mathcal{C}^0$ *in* $\mathcal{V}$*, and if the vector field is* $\mathcal{C}^k$ *(in space and parameters) it is* $\mathcal{C}^{k-1}$*.*

This result follows as an immediate rephrasing of the implicit function theorem (see [D]). A first consequence is that in a regular family of vector fields *a point attractor cannot disappear* as a function of the parameters at least until the spectrum of the tangent map $\mathrm{D}_2 X(\lambda, 0)$ hits the imaginary axis. If it does hit the imaginary axis, but not at the origin, the above theorem still applies but the stationary solution may become unstable. If the spectrum hits zero, the above theorem does not apply and the stationary solution may disappear. In other words, stationary solutions (stable or not) can only disappear when an eigenvalue of the tangent map crosses 0. By reversing the parametrization, we conclude also that it can only appear from nothing when an eigenvalue crosses 0.

We conclude from the above theorem that we will see curves of stationary solutions in phase space when the parameter is changed. It turns out that in order to study the geometry of these curves, the most useful parametrization is, in general, *not* the initial parameter λ. We will see later many instances of this phenomenon. This will be particularly important when we are dealing with the transition situations described earlier.

In order to illustrate this point we consider a $\mathcal{C}^\infty$ vector field depending in a $\mathcal{C}^\infty$ way on one real parameter λ, and with a stationary solution 0 for $\lambda = 0$. A natural question in view of the above discussion is to determine whether there is a curve of solutions passing through this point in the larger space whose coordinates are the parameters and the "spatial" coordinate in $\mathbf{R}^d$. That is to say, we are looking for a curve

$$\varepsilon \mapsto (\lambda(\varepsilon), M(\varepsilon))$$

such that $\lambda(0) = 0$, $M(0) = 0$, and $X(\lambda(\varepsilon), M(\varepsilon)) = 0$ when ε varies in a neighborhood $\mathcal{U}$ of 0. If we ask for a $\mathcal{C}^\infty$ curve, we can get a system of equations for the Taylor coefficients at $\varepsilon = 0$, and if this system can be solved, we get a formal power series for the solution curve. This system is easily obtained by differentiating the equation $X(\lambda(\varepsilon), M(\varepsilon)) = 0$ with respect to ε at the origin. Instead of discussing this problem in full generality we shall examine two simple but quite important examples, both in one dimension.

The first example is the vector field

$$X(\lambda, x) = -\lambda - x^2 \ .$$

The first derivative at 0 of the stationarity equation $-\lambda(\varepsilon) - x(\varepsilon)^2 = 0$ gives $\lambda'(0) = 0$, while the second derivative gives

$$\lambda''(0) + 2x'(0)^2 = 0 \ .$$

It is then easy to guess the solution $\lambda(\varepsilon) = -\varepsilon^2$, $x(\varepsilon) = \varepsilon$. We can now understand why at $\varepsilon = 0$, Theorem 10.1 does not apply. The curve $\varepsilon \mapsto (\lambda(\varepsilon), x(\varepsilon))$ is folded in such a way that it has no projection on the half-space $\lambda > 0$. In other words, there is no solution near 0 for λ small and positive. This is a rather strong singularity in the parameter λ. However, in the parameter ε the curve is perfectly smooth. In Section 22, we will put this reparametrization into a more general context.

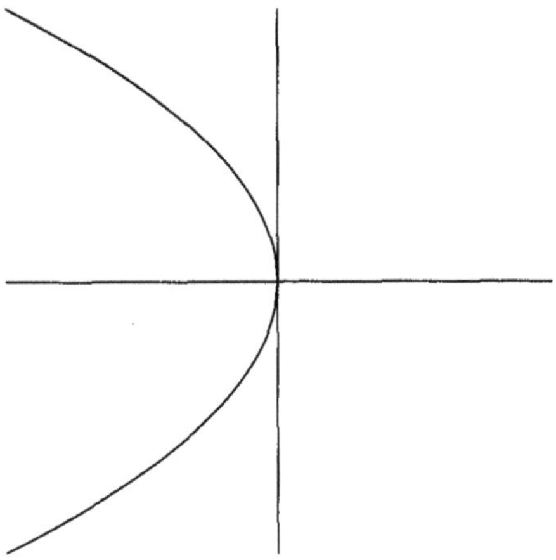

Fig. 4: A singularity appears in the bifurcation parameter λ; there are no solutions for $\lambda > 0$. On the other hand, the bifurcation curve itself is smooth.

The second example is given by $X(\lambda, x) = \lambda x - x^2$. We leave it to the reader to verify that we have two curves of solutions given by $\lambda(\varepsilon) = \varepsilon$, $x(\varepsilon) = 0$ and $\lambda(\varepsilon) = \varepsilon$, $x(\varepsilon) = \varepsilon$. These two curves cross at the origin, but otherwise they are regular.

So far, we have analyzed attractors as fixed geometrical objects. We now show how to generalize the reparametrization so as to capture not only the attractor, but also the *time scale* of the motion as a function of the parameter. We choose the time scale in the form

$$s = \vartheta_\varepsilon t \ .$$

Here ϑ_ε is a suitable C^∞ function (power series) in the parameter ε. We do the calculation in detail for the first of the examples above. We consider the problem

$$\frac{dx_\varepsilon(t)}{dt} = -\lambda_\varepsilon - x_\varepsilon^2(t) \ .$$

(We write now the parameter as an index, and derivatives with respect to the parameter as primes.) Setting $y_\varepsilon(\vartheta_\varepsilon t) = x_\varepsilon(t)$, the above equation becomes

$$\vartheta_\varepsilon \frac{dy_\varepsilon(t)}{dt} = -\lambda_\varepsilon - y_\varepsilon^2(t) \ . \tag{10.1}$$

We set $\lambda_0 = 0$ and $y_0(t) \equiv 0$. When $\lambda = 0$ and y is constant in time, *the time scale of motion is infinite*. This phenomenon is sometimes called critical slowing down. In order to extract interesting behavior at $\varepsilon = 0$, we set $\vartheta_0 = 0$. Differentiating (10.1) with respect to ε, we get, at $\varepsilon = 0$,

$$\vartheta_0' \frac{dy_0(t)}{dt} + \vartheta_0 \cdot \partial_\varepsilon \frac{dy_0(t)}{dt} = -\lambda_0' - 2y_0(t)y_0'(t) .$$

Since $y_0(t) \equiv 0$, the only remaining term is $0 = -\lambda_0'$ which is the solution we found before in the time-independent analysis. The second derivative yields, at $\varepsilon = 0$,

$$2\vartheta_0' \partial_\varepsilon \frac{dy_\varepsilon(t)}{dt}\Big|_{\varepsilon=0} = -\lambda_0'' - 2\big(y_0'(t)\big)^2 .$$

This equation has a regular solution, and its higher Taylor coefficients in ε can be derived recursively.

The fully general formulation of the above technique is left to the reader. We observe that letting the parameter ε tend to 0 after a correct scaling of the differential equation will in general lead to a nontrivial limiting equation.

The Theorem 10.1 does not apply to bifurcations from a stationary solution to a nonstationary solution (see for example the Hopf bifurcation described in Section 11). In the more general context of partial differential equations this situation of nonstationarity will be very frequent.

11. Bifurcation from a Point Attractor

We have seen in the preceding section that a point attractor of a vector field cannot disappear as parameters are changed unless the spectrum of the differential crosses the imaginary axis. The natural question is therefore what happens to the attractor and the dynamical behavior beyond this point. This is (one of) the goals of bifurcation theory, a rather large sum of results that can be arranged along several lines of thought. A value of the parameter where the attractor changes is called a bifurcation point.

We shall fix the notation as follows. We assume that for $\lambda < 0$, the vector field $X(\lambda, \cdot)$, $X : \mathbf{R} \times \mathbf{R}^d \to \mathbf{R}^d$ has a stable stationary solution M_λ. We shall simplify the notation by assuming that $M_0 = 0$. We shall also assume that we are at a bifurcation point, namely that the real matrix

$$\mathrm{D}_2 X(0, 0)$$

has its spectrum in the closed left half-plane with at least one point on the imaginary axis. We recall that the spectrum is composed of real eigenvalues and complex conjugated pairs.

An astonishing fact about bifurcation theory is the large (infinite) number of possible situations. To give a simple (but important) example, one can consider various possibilities for the complex eigenvalues of the matrix

$$\mathrm{D}_2 X(0, M_0) .$$

For large dimension d, one can have, for example, several pairs of conjugated imaginary eigenvalues appearing simultaneously at some parameter value (say, $\lambda = 0$, to be specific). Intuitively, each different pattern of imaginary eigenvalues can lead to different dynamical behavior for λ positive and small. Therefore, we need at this point some method to classify the different possibilities.

Before explaining this method, we want to argue that most of the complicated possibilities are unlikely from a physical point of view. We discuss this concept using as an example the following vector field:

$$X(\lambda, (x, y, z)) = \begin{pmatrix} \lambda x + y - x(x^2 + y^2 + z^2) \\ -x + \lambda y - y(x^2 + y^2 + z^2) \\ \lambda z - z(x^2 + y^2 + z^2) \end{pmatrix} .$$

For $\lambda < 0$, it is easy to verify that the origin in $\mathbf{R}^3$ is an attractor. In fact, it is a global attractor since

$$d(x^2 + y^2 + z^2)/dt = (\lambda - (x^2 + y^2 + z^2))(x^2 + y^2 + z^2) < \lambda(x^2 + y^2 + z^2) .$$

This implies that for any initial condition, the Euclidean norm of the solution tends to zero exponentially fast. For the differential, we have

$$\mathrm{D}_2 X(\lambda, (0,0,0)) = \begin{pmatrix} \lambda & 1 & 0 \\ -1 & \lambda & 0 \\ 0 & 0 & \lambda \end{pmatrix} .$$

Therefore, for $\lambda = 0$ the tangent matrix has an eigenvalue zero and a pair of complex conjugated eigenvalues $\pm i$. This degeneracy is, however, not stable under small perturbations. Consider for example the two parameter family of vector fields

$$Y(\lambda, \varepsilon, (x, y, z)) = X(\lambda, (x, y, z)) + \varepsilon \begin{pmatrix} 0 \\ 0 \\ z \end{pmatrix} .$$

For $\varepsilon = 0$ we obtain the previous vector field. For any ε, the origin is a stationary solution, and the differential is given by

$$\mathrm{D}_3 Y(\lambda, \varepsilon, (0,0,0)) = \begin{pmatrix} \lambda & 1 & 0 \\ -1 & \lambda & 0 \\ 0 & 0 & \lambda + \varepsilon \end{pmatrix} .$$

We see that if $\varepsilon < 0$, the bifurcation point is $\lambda = 0$ but it now corresponds only to a pair of complex eigenvalues crossing the imaginary axis. However, for $\varepsilon > 0$, the bifurcation point is $\lambda = -\varepsilon$ and it corresponds to a real eigenvalue crossing zero. In other words, the degenerate situation corresponding to $\varepsilon = 0$ is not stable, it can be destroyed by a small perturbation of the vector field.

From a physical point of view we shall be interested mostly in bifurcations which are stable under small perturbations of the vector field, so that they are more robust under a slightly wrong formulation of the physical principles. But it must be kept in mind that nonstable situations can (and do) occur in some circumstances which correspond to problems in physics.

At this point, the main question is therefore: What are the possible stable bifurcations? It can be shown that there are only two, one corresponds to a zero eigenvalue, the other one to a complex conjugate pair. We now formulate the associated bifurcation theorems which describe the (local) change of the dynamics when the parameter crosses the bifurcation point. We want to emphasize that these theorems are the ones to work with first: In the presence of a bifurcation of a point attractor one should first check if the hypotheses of these general theorems are satisfied. The reason is, of course, that these stable bifurcations are the most likely to occur. We shall indicate later some techniques that have been developed to deal with the situations where one cannot apply one of the following theorems.

The first bifurcation theorem is concerned with the so-called saddle node bifurcation.

Theorem 11.1. *Let $X(\lambda, \cdot)$ be a $\mathcal{C}^2$ vector field which is $\mathcal{C}^1$ in the parameter λ. Assume that for $\lambda < 0$ there is a stable stationary solution which is at the origin for $\lambda = 0$. Assume also that the matrix*

$$D_2X(0,0)$$

has an eigenvalue zero with eigenvector v and left eigenvector w, and the rest of the spectrum is in the open left half-plane. Assume also that

$$w(\partial_\lambda X)(0,0) \neq 0$$

and

$$\frac{wD_2^2X(0,0)[v,v]}{w(\partial_\lambda X)(0,0)} > 0 ,$$

when $v \neq 0$. Then, for $\lambda > 0$ small enough, there is no stationary solution near the origin. More precisely, for $\lambda < 0$ small enough, there is near the origin a nonstable stationary solution which collides with the stable stationary solution at $\lambda = 0$.

Remark. The notation $D_2^2X(0,0)[v_1, v_2]$ means the second (Fréchet) derivative of X with respect to its second argument, in the directions v_1 and v_2.

A typical example is given by the family of one dimensional differential equations

$$dx(t)/dt = -\lambda - x^2(t) .$$

We shall see later that the general case reduces in some sense to this quite elementary situation.

The Fig. 5 is a drawing in the space $\mathbf{R} \times \mathbf{R}^d$, with $(d = 1)$. It shows the locus of the stationary points versus the parameter. Such figures are called *bifurcation diagrams.*

The second main bifurcation theorem for attractors is associated with the so-called Hopf bifurcation.

Theorem 11.2. *Let $X(\lambda, \cdot)$ be a $\mathcal{C}^2$ vector field which is $\mathcal{C}^1$ in the parameter λ. Assume that for $\lambda < 0$ there is a stable stationary*

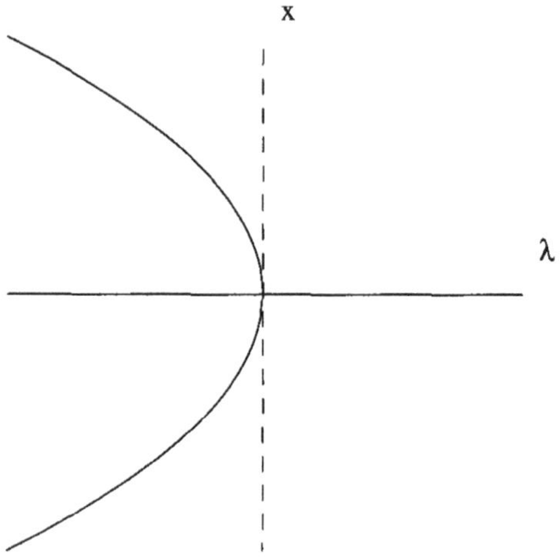

Fig. 5: The bifurcation diagram for the equation $\dot{x} = -\lambda - x^2$. The full lines show the set of stationary points for this equation in the x, λ plane.

solution which is at the origin for $\lambda = 0$. Assume also that the matrix

$$D_2X(0,0)$$

has two complex conjugated eigenvalues $\pm i\omega \neq 0$ and the rest of the spectrum in the open left half-plane. It follows that the eigenvalues $\Omega(\lambda)$ and $\overline{\Omega(\lambda)}$ which are purely imaginary at $\lambda = 0$ are differentiable in λ near $\lambda = 0$. Assume also that

$$\frac{d}{d\lambda}(\mathrm{Re}\,\Omega)(0) \neq 0 \ .$$

Generically, there is, for small $\lambda > 0$ or $\lambda < 0$, near zero a stationary solution and a closed invariant curve of radius $\mathcal{O}(\lambda^{1/2})$. These two invariant sets collapse to zero when λ goes to zero.

Remark. One can decide by a perturbative calculation to third order whether one is in the generic case and if the closed invariant curves occur for $\lambda > 0$ or $\lambda < 0$. This calculation will also show if the bifurcating circle is an attractor or a repeller. These calculations are lengthy, cf. [GH].

A typical example is given by the family of two coupled differential equations

$$dx(t)/dt = \lambda x + y - x(x^2 + y^2) ,$$
$$dy(t)/dt = -x + \lambda y - y(x^2 + y^2) ,$$

to which the case of a Hopf bifurcation leading to an attractor can be essentially reduced. This case is called the supercritical Hopf bifurcation.

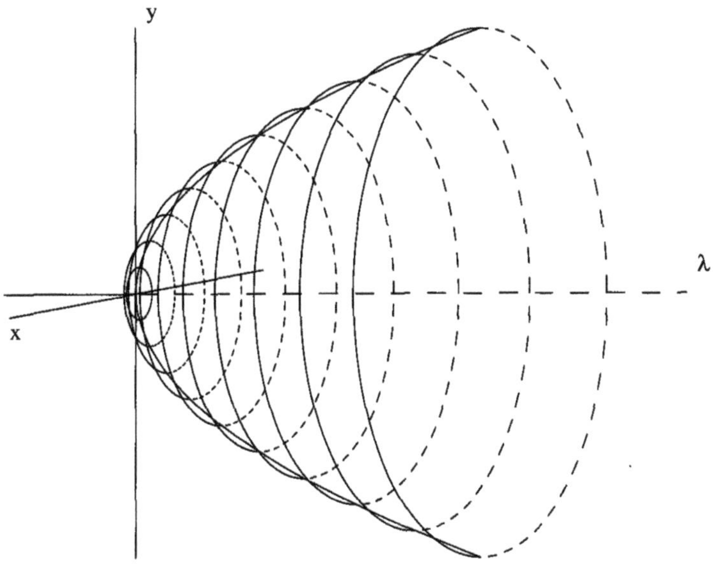

Fig. 6: A bifurcation diagram for a supercritical Hopf bifurcation. The x and y axes represent phase space, the λ axis is the parameter space. The invariant circles and the negative λ axis are *stable* in this case.

As explained before, the above bifurcations are the most likely ones to occur. It turns out, however, that one can meet other bifurcations. In our study of instabilities one of these more special types will occur. A simple and very common example is the so-called *transcritical bifurcation* which we illustrate by the following family of one dimensional differential equations

$$dx/dt = \lambda x - x^2 .$$

For $\lambda < 0$, the origin is an attractor. There is another stationary solution $x = \lambda$ which is a repeller. For $\lambda > 0$, the origin, which is still a stationary solution, is a repeller. The other stationary solution $x = \lambda$ is now the attractor. In other words, the two stationary solutions $x = 0$ and $x = \lambda$ have *exchanged* their stability. It is easy to verify that this is not a stable bifurcation. By adding a constant ε to the vector field one can verify that for $\varepsilon > 0$ one has two saddle node bifurcations and for $\varepsilon < 0$ no bifurcation at all.

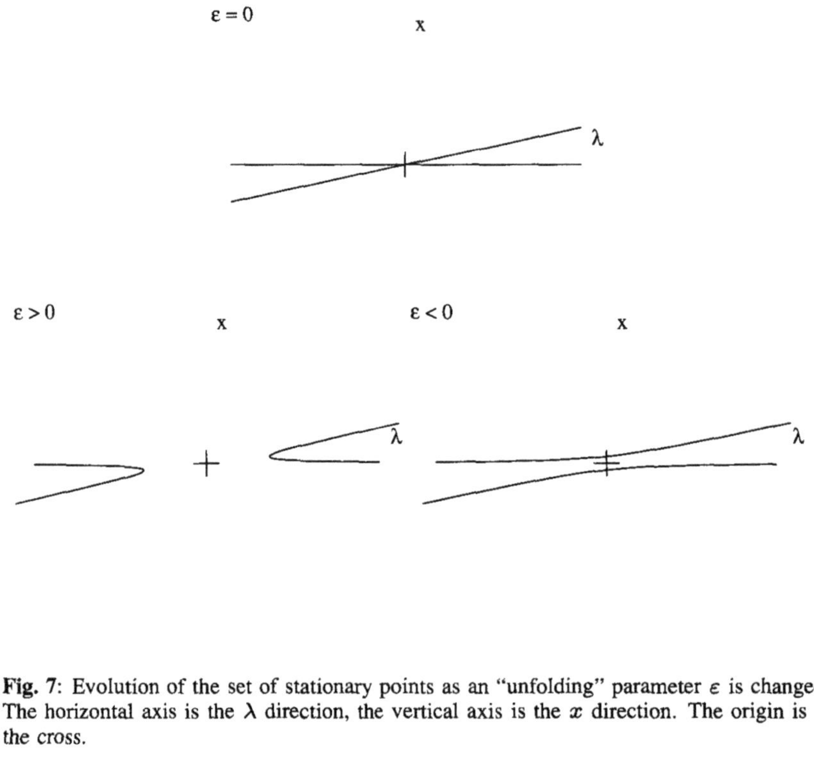

Fig. 7: Evolution of the set of stationary points as an "unfolding" parameter ε is changed. The horizontal axis is the λ direction, the vertical axis is the x direction. The origin is at the cross.

An important occurrence of this bifurcation is the well-known ε expansion for the study of second order phase transitions in Statistical Mechanics. The bifurcation parameter is usually denoted by ε and is equal to $4 - d$ where d is the spatial dimension of the system. There is a trivial fixed point for any ε with one unstable direction (temperature) for $\varepsilon < 0$

and two for $\varepsilon > 0$. (There is an additional unstable direction which can be eliminated by a change of scale.) There is also a nontrivial fixed point with one unstable direction which emerges from the trivial one for $\varepsilon > 0$ (smaller than 2).

Another simple nongeneric bifurcation is given by the family

$$dx/dt = \lambda x - x^3 .$$

For $\lambda < 0$ the origin is a global attractor. For $\lambda > 0$ the origin is still a stationary solution but it is a repeller. There are two other stable stationary solutions $x = \pm\sqrt{\lambda}$ which have emerged from the origin. This is called a *pitchfork bifurcation* because of the form of the bifurcation diagram.

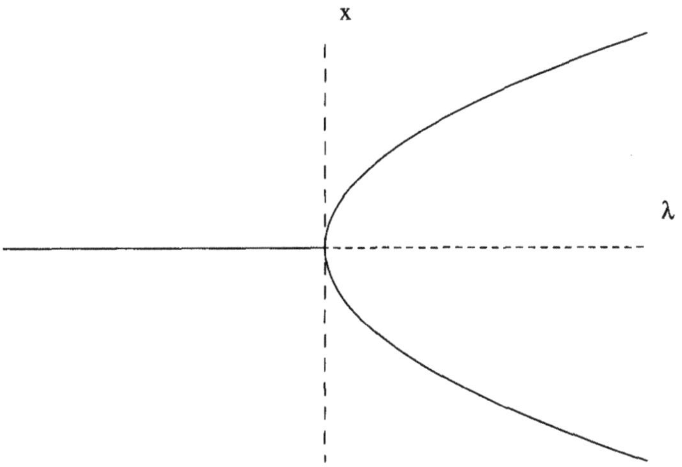

Fig. 8: The bifurcation diagram for a pitchfork bifurcation. The vertical axis is not part of the set of stationary points. The full lines are stable stationary points, and the dotted (horizontal) line is the set of unstable stationary points.

A simple example for which this type of bifurcation occurs is the two dimensional Hopf bifurcation discussed above. If we look at the evolution equation in polar coordinates, the radius

$$\rho = \left(x^2 + y^2\right)^{1/2}$$

has an evolution equation exactly as above. Of course, for $\lambda > 0$ we should only keep the positive solution $\rho = \sqrt{\lambda}$ which corresponds to a circle.

So far, we described only the case where a point attractor was bifurcating to some other attractor in a *continuous* way. More general interactions between point attractors and other invariant sets can occur when the parameters are changed. The case of the saddle node bifurcation already implies some discontinuity in the behavior of the attractor. We briefly describe some simple (but frequent) examples. We give only an informal discussion, referring the reader to the technical literature for the relevant theorems.

The first example is the inverse saddle node bifurcation which consists of the simultaneous appearance of a stable and an unstable stationary solution. A typical example is, of course, given by the one dimensional one parameter family of vector fields

$$dx/dt = \lambda - x^2 .$$

There is no attractor for $\lambda < 0$, an attractor ($x = \sqrt{\lambda}$) and a repeller ($x = -\sqrt{\lambda}$) for $\lambda > 0$.

Another important example is the so-called subcritical Hopf bifurcation (the Hopf bifurcation described before is sometimes called the supercritical Hopf bifurcation).

It occurs when an unstable limit cycle collapses with a stable stationary point to produce an unstable stationary point. A simple two dimensional example is given by

$$\begin{aligned} dx/dt &= \lambda x + y + x(x^2 + y^2) \\ dy/dt &= -x + \lambda y + y(x^2 + y^2) \end{aligned}$$

which has a subcritical Hopf bifurcation at $\lambda = 0$.

More complicated situations can, of course, occur. We describe only one which presents the phenomenon of hysteresis in the parameter. This is produced by an interaction between a subcritical pitchfork bifurcation and an inverse saddle node bifurcation. Consider the one dimensional vector field

$$dx/dt = \lambda x + x^3 - x^5 .$$

$x = 0$ is a stationary point which is stable for $\lambda < 0$ and unstable for $\lambda > 0$. For $\lambda = -1/4$ we have also two simultaneous inverse saddle node bifurcations at $x = \pm 1/\sqrt{2}$. Each gives rise to a branch of stable stationary solutions $\left(x = \pm\sqrt{\frac{1+\sqrt{1+4\lambda}}{2}}\right)$, and an unstable branch

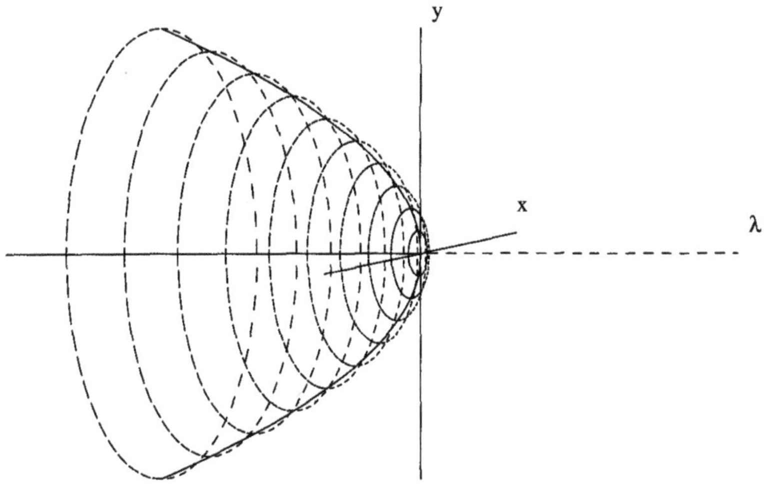

Fig. 9: A bifurcation diagram for a subcritical Hopf bifurcation. The x and y axes represent phase space, the λ axis is the parameter space. The invariant circles and the positive λ axis are *unstable* in this case.

$\left(x = \pm\sqrt{\frac{1-\sqrt{1+4\lambda}}{2}}\right)$. This last branch gives rise to the inverse pitchfork bifurcation of the origin at $\lambda = 0$.

We now explain why this example shows hysteresis. Imagine that we change the parameter λ very slowly (adiabatically). Starting from a value of λ smaller than $-1/4$, the variable x will be at the origin which is then the only attractor. Increasing λ, x will remain at the origin until we cross $\lambda = 0$. Beyond this point it could in principle stay at the origin which is a fixed solution. However, small perturbations of physical origin will move the system slightly out of the origin and it will then relax to one of the attractors $\left(x = \pm\sqrt{\frac{1+\sqrt{1+4\lambda}}{2}}\right)$ which is at a finite distance away. Suppose now that we decrease adiabatically λ. The system will follow the nontrivial branch of attractor until $\lambda = -1/4$ where it will suddenly jump back to the origin. We leave more complicated examples to the reader's imagination.

Another important source of new types of bifurcations is the presence of symmetries for the vector field. We refer to [R1] for a discussion of this situation.

12. Normal Forms

As we have seen before, some peculiarities in bifurcation problems are avoidable by small perturbations of one parameter families. A simple example was the simultaneous occurrence at the bifurcation point of an eigenvalue 0 and two complex conjugate imaginary eigenvalues. Although this phenomenon is exceptional for one parameter families of vector fields it is not exceptional for two parameter families. This leads to the definition of the codimension of a bifurcation. A bifurcation is said to be of codimension p if it occurs stably in families with p parameters (and unstably in families with $p-1$ parameters). Re-expressed in this terminology, we have classified above the codimension one bifurcations for vector fields, and listed a few codimension 2 bifurcations.

In order to classify the different possibilities more precisely, one has to introduce an equivalence relation. The notion goes back to Poincaré, and is very natural from a physical point of view. It is based on the idea that changes of variables are not going to modify the topological behavior of the dynamics.

Definition 12.1. *Two vector fields are topologically equivalent if there is a homeomorphism which conjugates the two associated flows.*

There is a weaker notion of orbital equivalence where one only asks that the homeomorphism transforms trajectories into trajectories preserving the direction of motion. This weaker notion of equivalence allows thus for reparametrizations of the time axis (which may depend on the point in phase space).

The simplest result associated to the topological equivalence is the Hartman-Grobman Theorem. To formulate it, we need a definition:

Definition 12.2. *A stationary solution of a C^1 vector field is called hyperbolic stationary solution if the spectrum of the linearization at this point does not intersect the imaginary axis.*

Theorem 12.3. *(Hartman-Grobman)[Ha]. A C^1 vector field is topologically equivalent to its linear part in a neighborhood of a hyperbolic stationary solution.*

Note that this beautiful result is not very useful for our present discussion since we are interested in bifurcation points which are nonhyperbolic by definition. It illustrates, however, the idea of topological equivalence: One would like to find inside each equivalence class a representative for which the dynamics is simple to study. Then, modulo the equivalence, all other elements of the class will have the same dynamics.

In what follows, we will only talk about a fixed vector field. Using, however, the trick of considering the trivial evolution equation for the parameter(s) we can treat in a similar way the case of families of vector fields provided that we impose on the change of variables that the part corresponding to the parameter does not depend on space.

The notion of topological equivalence defined above is very weak and is not well adapted to explicit computations. It is also not suited for quantitative discussions. To be in a more constructive situation, one has to look for more stringent *differentiable* equivalences. The following definition is well adapted to perturbative calculations. We shall say that two C^∞ vector fields X_1 and X_2 are formally equivalent if there is a formal power series ϕ which is a diffeomorphism* – in the sense of formal power series – such that

$$\big(\partial_x \phi(x)\big) \cdot X_1(x) = X_2(\phi(x)) . \tag{12.1}$$

(Recall that $\partial_x \phi$ is the matrix of derivatives.) If one can solve (12.1), then the two flows will be conjugated by a change of variables in phase space. The program of looking for a simpler representative in an equivalence class is usually attempted using this formal equivalence. Starting with a given vector field X one looks for a formal power series which simplifies the vector field as much as possible. In performing these computations one meets the interesting and important phenomenon of resonances.

* A differentiable map with a differentiable inverse.

12.1. Resonances

Assume that the vector field $X : \mathbf{R}^d \to \mathbf{R}^d$ has a stationary solution and let $\lambda_1, \dots, \lambda_d$ be the (possibly complex) eigenvalues of the differential of the vector field at the stationary solution.

Definition 12.4. *We say that the stationary solution has a* **resonance** *if there exists a choice of integers p_l (zero allowed) such that for at least one j one has*

$$\lambda_j = \sum_{l \neq j} p_l \lambda_l \ .$$

One says that the stationary solution has **small divisors** if, for all choices of p_l, the numbers

$$|\lambda_j - \sum_{l \neq j} p_l \lambda_l|$$

are different from zero but go rapidly to zero as $\sum_{l \neq j} |p_l|$ goes to ∞.

We shall now explain how resonances occur in conjugacy problems. Let X_1 and X_2 be two C^∞ vector fields in $\mathbf{R}^d$ with the origin as a stationary solution. We can write the two formal power series associated to these two vector fields and solve recursively for the conjugating formal power series. We shall denote by ϕ_n the sum of the homogeneous terms of degree n in the formal power series ϕ. In more detail, let $x_1, \dots, x_d$ denote the components of the vector x. Then, $\phi(x) = \sum_{n=0}^{\infty} \phi_n(x)$, with

$$\phi_n(x) = \sum_{\substack{i_1, \dots, i_d \geq 0 \\ i_1 + \dots + i_d = n}} \phi_n^{(i_1, \dots, i_d)} x_1^{i_1} \cdots x_d^{i_d} \ ,$$

where the coefficients $\phi_n^{(i_1, \dots, i_d)}$ are vectors in $\mathbf{R}^d$ (or $\mathbf{C}^d$). We shall obtain an infinite sequence of equations for the unknowns $(\phi_n)_{n \geq 0}$ by considering the terms of various order in the above conjugation equation (12.1). Since both vector fields are zero at the origin we must first have $\phi_0 = 0$. The equation for ϕ_1, which is a linear operator, is more interesting. We get, from the chain rule of differentiation,

$$\mathrm{D}\phi_1 X_1(0) = \mathrm{D}X_2(0)\phi_1 \ .$$

(Here, D is the derivative.) Recall now that if ϕ is the formal power series of a diffeomorphism, then ϕ_1, which is the differential at the origin, must be invertible. Therefore, we conclude that the two matrices $DX_1(0)$ and $DX_2(0)$ must be conjugate, otherwise the problem has no solution. In particular their spectrum must be the same, and according to the above definition they will have the same resonances. If this is not the case, the two vector fields are not even $\mathcal{C}^1$ conjugate.

So let us now assume that the two differential matrices are conjugate. Then we can use a linear change of variables for which they become equal. Assume we have done this transformation, and let A denote the common differential matrix at the origin. We can now write easily a set of recursive equations for our successive unknowns $(\phi_n)_{n\geq 0}$. We get

$$D\phi_n(x)Ax = A\phi_n(x) + \mathcal{F}_n(x) ,$$

where $\mathcal{F}_n$ is a quantity which depends on the formal power series of the two vector fields up to order n and on the functions ϕ_p for $p < n$. The recursiveness of the set of equations is now explicit. We get the equations, for each component j of the vector in $\mathbf{R}^d$,

$$\left(\sum_{\ell=1}^{d} i_\ell \lambda_\ell - \lambda_j\right) \phi_{n,j}^{(i_1,\dots,i_d)} = \mathcal{F}_{n,j}^{(i_1,\dots,i_d)} ,$$

which is solved by a simple division if we do not have a resonance. If we do have a resonance, the equation can only be solved if the right-hand side is zero. This means that the nonlinearity must be of a special form. If the right-hand side is zero, then there is a solution, but it is not unique. *In a certain sense, the appearance of a bifurcation parameter will be caused by this nonuniqueness.* (This statement holds only in the presence of one resonance.) We also note that the size of the solution can become large if we have small denominators. Infinitely many such large coefficients can eventually destroy the summability of the formal power series ϕ.

From a practical point of view, one tries to replace a vector field by an equivalent one as simple as possible. The simplest case would, of course, be to replace a vector field with the origin as a stationary solution by the linear vector field corresponding to the differential at the origin. We have seen that due to the resonances, this is, in general, not possible. There is, on the other hand, an interesting important result of Belitskii [Be] which says that one can conjugate any vector field to another one where only resonant terms remain. This conjugacy is even better than a formal power series, it is a regular diffeomorphism. Since there are, in general, infinitely many resonant terms, this result is satisfactory from a theoretical point of view, but not very useful for applications. A more realistic method is to start the formal reduction, and then to truncate the formal power series for the simplified vector field at some finite order. One then looks for the smallest order such that the truncated vector field belongs to the topological equivalence class of the original vector field. The vector field obtained in that way is called a normal form. We refer to [ET] for an efficient method of computing normal forms.

We refer to Guckenheimer-Holmes [GH], Arnold [Ar1] for more details and lists of codimension two normal forms.

13. Flows and Maps

13.1. Poincaré Sections

In the previous paragraphs we have discussed in some detail the problems of stability and bifurcation of point attractors. We have seen that in certain cases (Hopf bifurcation) they can bifurcate to limit cycles. We indicate how to discuss the stability of these limit cycles. The main tool was introduced by Poincaré and is called a Poincaré section. In this method, one constructs a Poincaré map also called a return map.

Let X be a regular (that is, sufficiently smooth) vector field with a regular invariant curve $\mathcal{C}$. In order to analyze the stability of $\mathcal{C}$, we consider the evolution of initial conditions close to $\mathcal{C}$. A convenient way to do this is to fix a point M on $\mathcal{C}$ and consider a hyperplane $\mathcal{P}$ transversal to $\mathcal{C}$ at M (for example the normal hyperplane). We now define a map $\mathcal{T}$ in a neighborhood $\mathcal{U}$ of M in $\mathcal{P}$ as follows. For any point P in $\mathcal{U}$, we consider the associated orbit. By continuity, this orbit will follow $\mathcal{C}$ for some time and if $\mathcal{U}$ is small it will cross $\mathcal{P}$ again near M (not necessarily in $\mathcal{U}$). We consider the first such crossing for which the vector field points into the same half-space as in M (by continuity, this is well defined if $\mathcal{U}$ is sufficiently small). The crossing point is called the image $\mathcal{T}(P)$ of P by the Poincaré map $\mathcal{T}$. Note that M is a fixed point of $\mathcal{T}$. It follows at once from the definitions that $\mathcal{C}$ is an attractor for X if and only if M is an attractor for $\mathcal{T}$.

At this point, the analysis of point attractors for vector fields can be generalized to an analysis of point attractors for maps. We shall only give the main ideas, referring the reader to the literature for proofs and details. We shall also only consider regular maps, that is, as differentiable

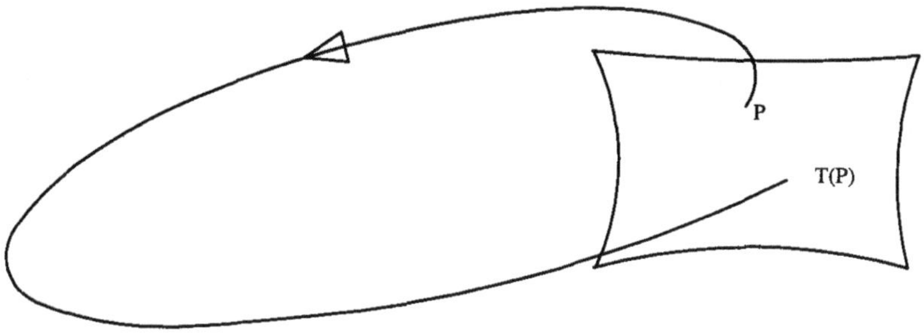

Fig. 10: The construction of the Poincaré map $P \longrightarrow \mathcal{T}(P)$.

as necessary.

Note that there is another way to associate a map to a vector field. It consists simply in considering the time one map. The construction of the Poincaré map has, however, the advantage of reducing the dimension of phase space by 1.

The first result is the condition for a fixed point to be an attractor.

Theorem 13.1. *If the spectrum of the differential of the regular map $\mathcal{T}$ at the fixed point M is contained in the open unit disk, then M is an attracting fixed point.*

Note also that since (in our case) $\mathcal{T}$ is a real map, the spectrum of its differential consists of real eigenvalues or of complex conjugated pairs.

We also have a stability theorem in case the above spectrum is contained in the open unit disk.

Theorem 13.2. *Let $\mathcal{T}_\lambda$ be a family of maps which depend C^1 on some parameter(s) λ. Assume also that for some parameter λ_0 the map $\mathcal{T}_{\lambda_0}$ has a fixed point M_{λ_0} which is an attractor such that the spectrum of $\mathrm{D}\mathcal{T}_{\lambda_0}(M_{\lambda_0})$ is inside the unit disk. Then there is a neighborhood of λ_0 where the same property holds.*

We conclude that changes of stability for point attractors of maps can only occur when the spectrum of the tangent map reaches the unit circle. As in the case of vector fields one would like to classify the different possible bifurcations. We shall only describe the three generic bifurcations for 1-parameter families of maps, and leave to the reader the formulation of the associated bifurcation theorems.

Theorem 13.3. *There are three generic bifurcations for point attractors of maps.*

1) *An isolated eigenvalue crosses the unit circle at $+1$. Then a stable and an unstable fixed point annihilate each other, and no attractor is left in the neighborhood. This is called a saddle node bifurcation. A typical example is the one dimensional mapping*

$$x \mapsto x + x^2 + \lambda$$

where the bifurcation point is $\lambda = 0$.

2) *An isolated eigenvalue crosses the unit circle at -1. The stable fixed point becomes unstable, and a periodic orbit of period two appears. This is an attractor composed of two points M_1 and M_2 which are exchanged by the map and such that an initial condition near enough to M_1 (resp. M_2) will flip nearer and nearer to the two points. This is called a pitchfork bifurcation or doubling bifurcation. A typical example is the one dimensional mapping*

$$x \mapsto 1 - \lambda x^2$$

with bifurcation point $\lambda = 3/4$.

3) *A pair of complex conjugate eigenvalues crosses the unit circle. This is a Hopf bifurcation, but the analysis is more subtle than in the case of vector field due to possible resonances (leading in particular to the so called Arnold tongues). One has to assume first that the eigenvalues are not roots of unity of degree less than five (for roots of degree one or two, see the above cases). Then an invariant stable curve appears continuously as a new attractor while the fixed point becomes unstable. Near the bifurcation value of the parameter it is a small deformation of a circle.*

As in the case of vector fields one should also, of course, distinguish supercritical bifurcations and subcritical bifurcations.

Coming back to the bifurcation problem for limit cycles, we see that a saddle node bifurcation for the return map corresponds to two invariant curves collapsing (one stable and the other unstable). A pitchfork bifurcation for the Poincaré map corresponds to the appearance of a new invariant (attractive) curve which wraps around twice before closing. A Hopf bifurcation for the Poincaré map corresponds to the attractor becoming an imbedded torus of dimension 2. More bifurcations can occur as parameters are changed; eventually leading to chaotic behavior. This is, however, beyond the scope of the present book and we refer the interested reader to the literature.

14. Bifurcation from a Simple Eigenvalue

Bifurcation from a simple eigenvalue (in Banach spaces) is a very old, and well-known, subject. We repeat it here because it is not always realized which are the most convenient parameters for bifurcation. The main idea is beautifully, and definitively, explained in the paper [CR]. This theorem is a generalization of the transcritical bifurcation.

Before going into technical details, we shall explain their main content. We will start with 1-dimensional equations, and then generalize to n-dimensional equations. We will work exclusively with fixed point equations, not with vector fields. This will be sufficient for later developments in this book.

14.1. The 1-Dimensional Case

Consider the nonlinear equation $\mathcal{N}_\lambda(u) = 0$ in the real unknown u, in one dimension, depending on a parameter λ. We assume that

1) $u = 0$ is a solution of $\mathcal{N}_\lambda(u) = 0$ for all λ in a neighborhood of $\lambda = 0$.
2) Let $e_\lambda = \partial_u \mathcal{N}_\lambda(0)$. Then
   a) $e_\lambda = 0$, when $\lambda = 0$ (zero eigenvalue),

b) $\partial_\lambda e_\lambda \neq 0$, when $\lambda = 0$ (transversality).

Then there will be a bifurcation from the zero solution at $\lambda = 0$ in the following sense:

There is a one-parameter family of solutions

$$u(\varepsilon) = \varepsilon + \mathcal{O}(\varepsilon^2) ,$$

for the problem $\mathcal{N}_{\lambda(\varepsilon)}(u(\varepsilon)) = 0$, *for some suitable parameter choice* $\lambda = \lambda(\varepsilon)$.

The main subtlety of the above statement is that the *bifurcation parameter is* ε *and not* λ, where ε measures the "amplitude" of the solution in the direction of the eigenvector v_0 at criticality. In general, one is tempted to take λ as a parameter, and this is obviously a good choice if $\partial_\varepsilon \lambda(\varepsilon) \neq 0$ for $\lambda = 0$. However, it is inadequate when this derivative vanishes, as it happens for example in the case of a pitchfork bifurcation. The most trivial examples of this phenomenon are given as follows:

Example 1: The problem is of the form

$$\mathcal{N}_\lambda(u) = \lambda u + u^2 ,$$

with solutions $u = 0$, eigenvector $v_0 = 1$ and eigenvalue $e_\lambda = \lambda$. Then we have a family of bifurcating solutions of the form $u_\lambda = -\lambda$, where we can take λ as a parameter, but the more generic way is to write

$$\begin{aligned} \lambda(\varepsilon) &= \varepsilon , \\ u(\varepsilon) &= -\varepsilon . \end{aligned}$$

Example 2: The problem is of the form

$$\mathcal{N}_\lambda(u) = \lambda u - u^3 ,$$

with solutions $u = 0$, eigenvector $v_0 = 1$ and eigenvalue $e_\lambda = \lambda$. Then we have a family of bifurcating solutions of the form $u_\lambda = \pm\sqrt{\lambda}$, where we can take λ as a parameter, but now the advantages of the generic parametrization are obvious

$$\begin{aligned} \lambda(\varepsilon) &= \varepsilon^2 , \\ u(\varepsilon) &= \varepsilon . \end{aligned}$$

Example 3: The problem is of the form

$$\mathcal{N}_\lambda(u) = \lambda u \ ,$$

with solutions $u = 0$, eigenvector $v_\lambda = 1$ and eigenvalue $e_\lambda = \lambda$. This case may seem too trivial to the reader, but we will see later that it points in the right direction, because it shows that the amplitude is the relevant quantity. We have a family of bifurcating solutions of the form $u_\varepsilon = \varepsilon$, which *cannot be parametrized by* λ *since* λ *is always zero.* But in ε nothing dramatic happens:

$$\lambda(\varepsilon) = 0 \ ,$$
$$u(\varepsilon) = \varepsilon \ .$$

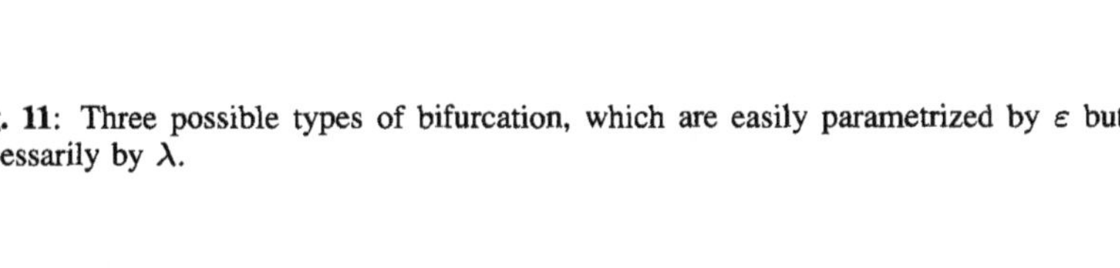

Fig. 11: Three possible types of bifurcation, which are easily parametrized by ε but not necessarily by λ.

Remark. Assume the problem is of the form $\mathcal{M}_\lambda(u) = 0$ and we know a solution $u_0(\lambda)$. Then we are reduced to the previous case by considering $\mathcal{N}_\lambda(u) = \mathcal{M}_\lambda(u_0(\lambda) + u)$. In other words, this is a change of origin in solution space.

14.2. The n-Dimensional Case

We consider the nonlinear equation $\mathcal{N}_\lambda(u) = 0$ where u varies in $\mathbf{X} = \mathbf{R}^n$, and $\mathcal{N}_\lambda$ takes values in $\mathbf{Y} = \mathbf{R}^n$. In the general case, we will allow $\mathbf{Y}$ to be different from $\mathbf{X}$. We assume that

1) $u = 0$ is a solution of $\mathcal{N}_\lambda(u) = 0$ for all λ in a neighborhood of $\lambda = 0$.
2) The derivative $\partial_u \mathcal{N}_\lambda$ evaluated at $u = 0$ is a symmetric $n \times n$ matrix with a simple, isolated eigenvalue e_λ and an eigenvector v_λ, satisfying
   a) $e_\lambda = 0$, when $\lambda = 0$ (zero eigenvalue),
   b) $\partial_\lambda \partial_u \mathcal{N}_\lambda(0) v_0$ is not orthogonal to v_0, when $\lambda = 0$ (transversality).

Remark. In the general statement below, we shall drop the symmetry requirement for $\partial_u \mathcal{N}_\lambda(0)$. Accordingly, we will have to replace the non-orthogonality statement 2b) by a more abstract transversality condition (cf. d) in the theorem below).

Under the above assumptions, there will be a bifurcation from the zero solution at $\lambda = 0$ in the following sense:

There is a one-parameter family of solutions

$$u(\varepsilon) = \varepsilon v_0 + \mathcal{O}(\varepsilon^2) ,$$

for the problem $\mathcal{N}_{\lambda(\varepsilon)}(u(\varepsilon)) = 0$, for some suitable parameter choice $\lambda = \lambda(\varepsilon)$. The function $\lambda(\varepsilon)$ is regular.

14.3. The Infinite-Dimensional Case

We are now ready for the statement of a general theorem, covering all of the cases above, and much more:

Theorem 14.1. *[CR] Let $\mathbf{X}, \mathbf{Y}$, be Banach spaces and let $\mathbf{V}$ be a neighborhood of the origin in $\mathbf{X}$. Consider a map*

$$\mathcal{N} : (-1, 1) \times \mathbf{V} \to \mathbf{Y} ,$$
$$\mathcal{N} : (\lambda, u) \mapsto \mathcal{N}_\lambda(u) ,$$

for which the derivatives $\partial_u \mathcal{N}$, $\partial_\lambda \mathcal{N}$, $\partial_\lambda \partial_u \mathcal{N}$ exist and are continuous. Assume that

a) $\mathcal{N}_\lambda(0) = 0$ for $|\lambda| < 1$,
b) The null space, $\mathbf{N}$, of $\partial_u \mathcal{N}_0(0)$ has dimension 1.
c) The range, $\mathbf{T}$, of $\partial_u \mathcal{N}_0(0)$ has codimension 1 in $\mathbf{Y}$.
d) Let $v_0 \neq 0$ be an element of $\mathbf{N}$. Then $\partial_\lambda \partial_u \mathcal{N}_0(0) v_0 \notin \mathbf{T}$.

Denote by $\mathbf{S}$ any supplement of $\mathbf{N}$ in $\mathbf{X}$. Then there are an interval $(-a, a) \subset (-1, 1)$ and two functions $\lambda : (-a, a) \to \mathbf{R}$ and $\eta : (-a, a) \to \mathbf{S}$ satisfying $\lambda(0) = 0$, $\eta(0) = 0$ such that

$$\mathcal{N}_{\lambda(\varepsilon)}(\varepsilon v_0 + \varepsilon \eta(\varepsilon)) = 0 .$$

The only other solutions to the problem $\mathcal{N}_\lambda(u) = 0$ in this neighborhood are $\lambda = \varepsilon$, $u = 0$. If $\partial_u \partial_u \mathcal{N}$ is also continuous in u and λ, then the functions $\lambda(\varepsilon)$ and $\eta(\varepsilon)$ are once continuously differentiable.

Remark. A supplement $\mathbf{S}$ of a subspace $\mathbf{N}$ in a space $\mathbf{X}$ is defined as follows: $\mathbf{S}$ is a subspace of $\mathbf{X}$ such that $\mathbf{S}$ and $\mathbf{N}$ span $\mathbf{X}$ and the dimension of $\mathbf{S}$ equals the codimension of $\mathbf{N}$ (in $\mathbf{X}$). Note that the supplement is, in general, not unique.

Remark. In the 1-dimensional example above, we have $\mathbf{X} = \mathbf{Y} = \mathbf{R}$, and $\mathbf{S}$ is empty. In the n-dimensional example, we have $\mathbf{X} = \mathbf{Y} = \mathbf{R}^n$, and $\mathbf{S}$ is the $n-1$-dimensional subspace in $\mathbf{R}^n$ which is orthogonal to the subspace spanned by the eigenvector v_0.

Proof. The proof is an application of the implicit function theorem. In order to transform the problem to a suitable form one defines a function

$$f(\varepsilon, \lambda, \eta) = \begin{cases} \varepsilon^{-1} \mathcal{N}_\lambda(\varepsilon v_0 + \varepsilon \eta) & \text{if } \varepsilon \neq 0, \\ \partial_u \mathcal{N}_\lambda(0)(v_0 + \eta) & \text{if } \varepsilon = 0. \end{cases} \tag{14.1}$$

Note that $f(0, \lambda, \eta)$ is linear in η. This function is defined for those $(\varepsilon, \lambda, \eta)$ for which $\varepsilon(v_0 + \eta) \in \mathbf{V}$, and $|\lambda| < 1$. We are asked to describe solutions of $f = 0$ in the form $f(\varepsilon, \lambda(\varepsilon), \eta(\varepsilon)) = 0$. The partial derivatives of f with respect to λ and η are continuous in $(\varepsilon, \lambda, \eta)$, and, by b), d), we have

$$f(0,0,0) = \partial_u \mathcal{N}_0(0) v_0 = 0 . \tag{14.2}$$

The Fréchet derivative of the map $(\lambda, \eta) \to f(0, \lambda, \eta)$ at $(\lambda, \eta) = (0, 0)$ is the linear map

$$(\delta\lambda,\ \delta\eta) \mapsto \delta\lambda \cdot \partial_\lambda \partial_u \mathcal{N}_0(0) v_0 \ + \ \partial_u \mathcal{N}_0(0)\, \delta\eta , \tag{14.3}$$

which maps $\mathbf{R} \times \mathbf{S}$ into $\mathbf{Y}$. The assumptions of the theorem imply that this map is an isomorphism onto $\mathbf{Y}$, that is, it is 1–1 and invertible, and therefore the implicit function theorem [D] implies the existence of functions $\lambda(\varepsilon)$ and $\eta(\varepsilon)$ with the required properties. It remains only to show that there are no other continuous solutions to the equation $\mathcal{N}_\lambda(u) = 0$ near $u = 0$ than those enumerated in the theorem. But by the implicit function theorem, we know that the solutions bifurcating into the direction of v_0 are unique. If we try to define f for solutions pointing into any other direction, $v_1 \neq 0$, $v_1 \notin \mathbf{N}$ by the analogue of (14.1) then the equation (14.2) will not be satisfied, by assumption b). Hence the solution curve in the direction v_1 cannot be continuous at 0.

Remark. The implicit function theorem, as applied here, gives a constructive method to find the perturbation series for λ and η as a function of ε, provided $\mathcal{N}$ is sufficiently differentiable. We adapt the standard formulae to the case at hand. Note that we are looking for a zero of the function f, as a power series in ε. To fix the notation, we recall the standard strategy: Assume F maps $\mathbf{R} \times \mathbf{X}'$ to $\mathbf{Y}'$

$$F : (\varepsilon, h) \mapsto F(\varepsilon, h) .$$

In our case, h is the pair λ, η and F is given by $F(\varepsilon, h) = \varepsilon^{-1}\mathcal{N}_\lambda(\varepsilon v_0 + \varepsilon\eta)$. One further assumes that $F(0,0) = 0$ and that $\partial_h F(0,0)$ is invertible. Expanding in ε and h we find

$$F(\varepsilon, h) = \varepsilon\partial_\varepsilon F(0,0) + \partial_h F(0,0)\, h + \mathcal{R}(\varepsilon, h) ,$$

with

$$\|\mathcal{R}(\varepsilon, h)\| \leq \text{const.} \left(|\varepsilon| + \|h\|\right)^2 .$$

Start now with $h_0 = 0$ and define h_{n+1} by

$$h_{n+1} = -\left(\partial_h F(0,0)\right)^{-1} \left(\varepsilon\partial_\varepsilon F(0,0) + \mathcal{R}(\varepsilon, h_n)\right) . \tag{14.4}$$

Then h_n will be correct to order n in ε. Applying this to our case, we find that $\partial_h F(0,0)$ is the linear operator (14.3), and $\varepsilon\partial_\varepsilon F(0,0)$ is the second order derivative

$$\varepsilon\partial_\varepsilon F(0,0) = \tfrac{1}{2}\partial_u\partial_u\mathcal{N}_0(0)[v_0, v_0] . \tag{14.5}$$

The notation in (14.5) is the generalization to a multidimensional situation of the following identity: If $F(\varepsilon, x) = \varepsilon^{-1}G(\varepsilon x)$, with $G(0) = 0$, then

$$\begin{aligned} F(0,0) &= (\partial_x G)(0) , \\ \varepsilon\partial_\varepsilon F(0,0) &= \tfrac{1}{2}(\partial_x^2 G)(0) \cdot x^2 . \end{aligned}$$

The only practical difficulty in applying (14.4) is the inversion of the matrix $\partial_h F(0,0)$. In finite dimension (or in the finite dimensional restriction of an infinite dimensional problem) this should be possible at least on a computer. Of course, in some cases, it may be possible to compute the spectrum and the eigenspaces explicitly.

Remark. An efficient algorithm to compute solutions (non-perturbatively) follows from the implicit function theorem: The solution can be found by the following variant of the Newton algorithm. This is similar to the perturbative calculation described above, but now the tangent map $\partial_h F$ is evaluated *at the current solution.* This means more computational work for one iteration step, but on the other hand, the convergence will be quadratic. The analogue of (14.4) (for $h_n = h_n(\varepsilon)$ with ε fixed) is

$$h_{n+1} = h_n - \left(\partial_h F(\varepsilon, h_n)\right)^{-1} F(\varepsilon, h_n) . \tag{14.6}$$

In terms of $\mathcal{N}$ this yields

$$(\lambda, \eta)_{n+1} = (\lambda, \eta)_n - (A_n)^{-1} \mathcal{N}_{\lambda_n}(\varepsilon v_0 + \varepsilon \eta_n) ,$$

where A_n is the matrix (linear map $\mathbf{R} \times \mathbf{V} \to \mathbf{Y}$)

$$(\delta\lambda,\ \delta\eta) \mapsto \delta\lambda \cdot \partial_\lambda \partial_u \mathcal{N}_{\lambda_n}(\varepsilon v_0 + \varepsilon \eta_n) v_0 + \partial_u \mathcal{N}_{\lambda_n}(\varepsilon v_0 + \varepsilon \eta_n)\, \delta\eta .$$

We state without proof an extension of Theorem 14.1 to several parameters (their number will be p).

Theorem 14.2. *Let* $\mathbf{X}, \mathbf{Y}$, *be Banach spaces and let* $\mathbf{V}$ *be a neighborhood of the origin in* $\mathbf{X}$. *Let* p *be an integer. Consider a map*

$$\mathcal{N} : (-1,\ 1)^p \times \mathbf{V} \to \mathbf{Y} ,$$
$$\mathcal{N} : (\lambda, u) \mapsto \mathcal{N}_\lambda(u) ,$$

for which the derivatives $\partial_u \mathcal{N}$, $\partial_\lambda \mathcal{N}$, $\partial_\lambda \partial_u \mathcal{N}$ *exist and are continuous. Assume that*

a) $\mathcal{N}_\lambda(0) = 0$ *for* $\lambda \in (-1, 1)^p$.

b) The null space, $\mathbf{N}$, *of* $\partial_u \mathcal{N}_0(0)$ *has dimension* p.

c) The range, $\mathbf{T}$, *of* $\partial_u \mathcal{N}_0(0)$ *has codimension* p *in* $\mathbf{Y}$.

d) There is a vector v_0 in **N** *such that the range* $\mathbf{T}_0$ *of the linear operator* $\partial_\lambda \partial_u \mathcal{N}_0(0)v_0$ *from* $\mathbf{R}^p$ *to* **Y** *has dimension* p *and satisfies* $\mathbf{T} \cap \mathbf{T}_0 = \{0\}$.

Denote by **S** *any supplement of* **N** *in* **X**. *Then there are an interval* $(-a, a) \subset (-1, 1)$ *and two functions* $\lambda : (-a, a) \to \mathbf{R}^p$ *and* $\eta : (-a, a) \to \mathbf{S}$ *satisfying* $\lambda(0) = 0$, $\eta(0) = 0$ *such that*

$$\mathcal{N}_{\lambda(\varepsilon)}(\varepsilon v_0 + \varepsilon\eta(\varepsilon)) = 0 .$$

If $\partial_u \partial_u \mathcal{N}$ *is continuous, then the functions* $\lambda(\varepsilon)$ *and* $\eta(\varepsilon)$ *are once continuously differentiable.*

CHAPTER IV
STATIONARY AND QUASISTATIONARY SOLUTIONS

15. Introduction

In Chapter III, we considered the time evolution of physical systems which have only a finite number of excited relevant degrees of freedom. The other degrees of freedom whose number may be infinite were, so to speak, slaves of the excited modes.

Now, we discuss systems in which an infinite number of degrees of freedom participate in the relevant physical evolution. In general, this occurs when we consider spatially infinitely extended systems. This is a situation which is reminiscent of Statistical Mechanics where one shows that large systems behave like infinite ones. Here, one can hope for a similar behavior, namely one can try to analyze large systems like infinite ones, correcting afterwards for the finiteness (for example by examining the influence of far away boundary conditions). Mathematicians and physicists have recognized the technical interest of this idea for computing spectra of linearized evolutions. The very dense discrete spectrum becomes continuous in the infinite domain and is therefore more accessible to Fourier analysis. Of course, the results obtained by this method are not as accurate since we are basically making some approximations. Another difficulty is that continuous spectra are more difficult to discuss than discrete ones. They are easily analyzed for differential operators with constant coefficients but become immediately more difficult when the coefficients are non-constant. We recall that on top of a continuous spectrum one can have a discrete one, and also that the nature of the spectrum can be quite strange (densely discrete, singular continuous etc.). Most of the time we shall be concerned with absolutely continuous essential spectrum but we have no doubt that many interesting physical effects involving more complicated spectral properties still have to be discovered. Although the description of the difficulties may sound somewhat pessimistic, we are convinced that the study of infinite systems is the right approach for the study of instabilities in large systems.

The state of an infinite system at a given time is described by a (vector valued) function u on $\mathbf{R}^d$ with values in some linear space $\mathbf{R}^\nu$ (mathematicians would, of course, consider functions from an unbounded manifold to another one, note that it may sometimes be useful to have the target space curved but we shall not do this here). We shall denote by $u(x,t)$ the state of the system at time t and spatial position x. We have seen in Chapter I examples where this (macroscopic) description of a state is employed. In hydrodynamics, $u(x,t)$ is the velocity of the fluid at time t and space position x. In chemical systems, $u(x,t)$ is the vector of concentrations of the various chemical species. Some other frequent examples include the temperature, various order parameters, electromagnetic fields etc.

In what follows, we will have to use functional analysis, and therefore we have to fix some function space for the state u at a fixed time. Here again, we shall base our assumptions on some physically reasonable ideas. First, the function u must be uniformly bounded in space (see the examples in Chapter I). It should also be regular (differentiable). There may be, of course, interesting exceptions to this last assumption; one of the most well-known is the theory of shock waves. We shall, however, not consider these situations in the present book. The natural family of function spaces emerging from physical constraints is given by a requirement that the function u and its derivatives are bounded for $x \in \mathbf{R}^d$. In some situations it will be interesting to introduce various spatial weights to discuss particular problems. For example for the study of fronts one has to introduce spaces of functions with a prescribed decay at infinity. We shall define these special spaces when needed.

We will see that it is an easy exercise to verify that spaces of the above type are Banach spaces. Moreover, if the weights are suitably chosen, these spaces form an algebra for the pointwise multiplication of the functions. Hence they are well fitted for the discussion of nonlinear problems. Note also that if the weights are reasonable, these function spaces are subspaces of the Schwartz space of tempered distributions, and therefore we can work with the Fourier transform of their elements.

There are also other interesting (less general) families of function spaces which satisfy the above physical requirements. We shall only mention the class of quasiperiodic functions. Although there are very few results concerning these functions in the present context, we want to mention that one can define sequences of function spaces of various regularity which are algebras (see for example [Sh] for a review).

The next question to discuss is the form of the time evolution. Again, from the physical examples considered above, we shall make some restrictions on the form of the time evolution. We will assume that there is a given nonlinear differential (or pseudo-differential operator) denoted by $\mathcal{N}$ which is quasilinear and elliptic. The term quasilinear means that the differential part of maximal order of the operator is *linear*, and the term elliptic means that the highest order derivative is positive definite.

All the equations we discussed in Chapter I are of this type. Ellipticity means that the underlying physical system is dissipative (in every point of phase space).

The time evolution is then given by the equation

$$\partial_t u = \mathcal{N}(u) . \tag{15.1}$$

One could also consider situations where the operator $\mathcal{N}$ depends on time. A natural example is the case of a periodic forcing. For simplicity we shall not consider systematically this more general situation. Note also that $\mathcal{N}$ may depend explicitly on the space variable. We will discuss problems of this type in Section 21.

We now discuss an important difference with the dynamical systems studied in Chapter III. It is related to the space-time geometry and is only relevant in infinite systems. The equation (15.1) is formulated in a special frame called the laboratory frame. If $u(x,t)$ is a solution of equation (15.1), it is often useful to consider this solution in a moving frame. In other words, we shall consider the point of view of some observer moving with respect to the laboratory. If the moving frame has a constant speed c (a vector if the space dimension d is larger than 1), one can write an equation for the observation u_c in that frame such that u_c corresponds to the solution u in the laboratory frame. There is a simple relation between u and u_c which is a consequence of the Galilean invariance of the system (we shall not consider relativistic problems here). For example if u is a scalar quantity, we have $u_c(x,t) = u(x+ct,t)$. More involved relations can also occur, depending on the physical interpretation of u. For example, if u is a velocity, we have $u_c(x,t) = u(x+ct,t) - c$. From these relations it is easy to derive the evolution equation for u_c. In the case of a scalar u, equation (15.1), when observed in a frame moving at speed c, becomes

$$\partial_t u_c(x,t) = \mathcal{N}(u_c)(x,t) - c \cdot \partial_x u_c(x,t) ,$$

where $\cdot$ denotes the scalar product.

One of the most productive assumptions of the theory is that $\mathcal{N}$ depends differentiably on a parameter hereafter denoted by α. Although this parameter may be a vector, we shall restrict ourselves to the case of a unique real number. A typical example is the viscosity in the Navier-Stokes equation. As we explained in Chapter III, varying a relevant parameter is important for the understanding of the dynamical behavior of a physical system both from an experimental and from a theoretical point of view.

Following the general methods for the study of dynamical systems which we have developed in Chapter III, we shall assume that one has been able to identify a region of parameter space where the asymptotic time evolution of the system (or at least of part of its phase space) is simple. In most cases (meaning in most cases studied up to now), the nonlinear differential operator $\mathcal{N}$ is homogeneous (that is, it does not depend explicitly on the space variables). In such cases, the simplest possible asymptotic time behavior is described by a spatially constant stationary solution, namely a function u constant in space and time which satisfies equation (15.1). For such special solutions, the evolution equation simplifies to

$$\mathcal{N}(u) = 0 .$$

An easy example is the motion of a fluid at a uniform speed. A less trivial example is the conducting state in the Rayleigh-Bénard experiment. The fluid is at rest but the temperature has a nonconstant linear profile.

We shall see that the definition of simple state given above is too stringent to be really interesting for applications. We shall therefore consider a more general definition of "simple" states which we call quasistationary solutions. With one exception to be discussed in Chapter VII, we use the following informal definition.

Definition. A solution of the equation

$$\partial_t u = \mathcal{N}(u)$$

is called **quasistationary** if it is time-independent either in the laboratory frame or in a frame moving with constant velocity.

Of course, one could imagine other interesting situations like time periodic spatially constant solutions, but we will not encounter such examples. A front, to be defined in Chapter VII, will be a solution of a more complicated nature than quasistationary solutions, but there are some simple fronts which are time-independent in a frame moving with fixed speed. (This is the exception to the above definition.)

16. Linear Stability Analysis

Before discussing (some of) the various possibilities for quasistationary time evolution, we shall describe a classification of the instabilities which is commonly used in the physics literature and is often attributed to Bers and Briggs in plasma physics. In order to maintain the exposition at a reasonably elementary technical level, we shall assume that the system (15.1) is homogeneous and has a spatially uniform stationary solution U.

As explained before, an instability is detected by looking at the spectrum of the linear evolution around U. If we have instability (part of the spectrum in the right half-plane), a small generic perturbation will grow exponentially fast. This implies that the nonlinear evolution of this perturbed initial condition will be controllable only for a finite amount of time, that is, until it has moved to a macroscopic distance from the stationary solution. Beyond this time, one has to deal with global nonlinear effects which may lead to a different asymptotic time behavior. Nevertheless, there is a finite initial time interval, valid until the amplitude has reached a certain size, where the linearized analysis is meaningful. In this section, we deal exclusively with this time interval, the remainder of this chapter will deal with the nonlinear effects, in a time-independent framework.

We shall denote by $A = \mathrm{D}\mathcal{N}_U$ the linearization of $\mathcal{N}$ at the stationary solution U, defined by

$$\varepsilon A v \equiv \mathcal{N}(U + \varepsilon v) - \mathcal{N}(U) + \mathcal{O}(\varepsilon^2) .$$

This is called the Fréchet derivative of $\mathcal{N}$ at U in the direction v. As explained above, we shall assume that A is a matrix valued differential operator with constant coefficients. The linearized evolution can now be formulated as follows. Given a (bounded) initial condition $v(x, 0)$, one wants to determine the asymptotic behavior of the function $v(x, t)$ satisfying

$$\partial_t v(x, t) = A v(x, t) .$$

Since we are dealing with partial differential operators with (matrix valued) constant coefficients, that is, since A does not depend explicitly on x, we can use Fourier analysis to solve the evolution problem. The partial differential operators become then matrix valued functions of the Fourier modes which can be triangularized (and eventually diagonalized). The integration of the time evolution is then a trivial exercise. To illustrate this procedure, we now consider the following rather typical example.

First, we set the space dimension to $d = 1$ and the order parameter dimension to $\nu = 1$. The evolution equation is given by the so-called nonlinear heat equation

$$\partial_t u = \partial_x^2 u + \alpha u - u^3 .$$

The function $U \equiv 0$ is a stationary solution. The linearized evolution around the stationary solution $U = 0$ is given by

$$\partial_t v = \partial_x^2 v + \alpha v .$$

Using Fourier transform with respect to the space variable x, it is easy to integrate explicitly the time evolution. One gets

$$v(\xi, t) = e^{t(\alpha - \xi^2)} v(\xi, 0) ,$$

where we have denoted by $v(\xi, t)$ the Fourier transform of the function $v(x, t)$. We immediately conclude that the stationary solution $U = 0$ is linearly stable for $\alpha < 0$ and unstable for $\alpha > 0$. If $\alpha > 0$, any initial function $v(\xi, 0)$ which is nonzero for some ξ satisfying $\xi^2 < \alpha$, will lead to a solution (of the linearized problem) which blows up as $t \to \infty$. The spectrum of the linearized operator is continuous, and is the set $\{\alpha - \xi^2 : \xi \in \mathbf{R}\}$.

16.1. Definition of Stability

Before we go on with further examples, we shall treat instabilities in a somewhat more general setting. Since we assume that A is a (matrix valued) differential operator with constant coefficients, its Fourier transform will be a (matrix valued) multiplication operator, depending on the parameter α and on the Fourier mode ξ. For fixed α, we consider the set of eigenvalues* $\sigma(\xi)$ of A as a function of ξ and call this set Σ_α, the spectrum of A (for that α). There are situations where $\sigma(\xi)$ is not a differentiable function of ξ, for example $\sigma(\xi) = -|\xi|^3$. We shall not consider such functions σ.

Definition. We shall say that the system

$$\partial_t u = \mathcal{N}_\alpha(u)$$

is *linearly stable, linearly unstable or marginal* depending on whether the spectrum Σ_α lies entirely in the open left half-space, intersects the right half-space or lies in the closed left half-space and contains points on the imaginary axis. See Fig. 12.

* There can be several eigenvalues for one ξ, but we will not, in this section, distinguish them with an index.

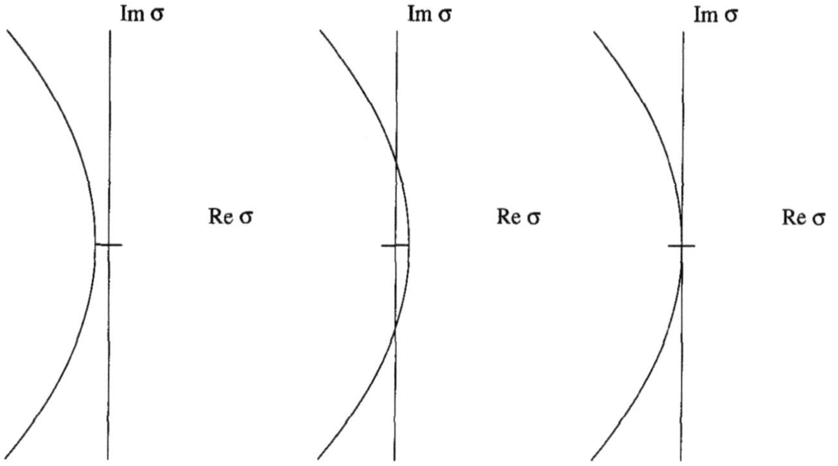

Fig. 12: Three (local) possibilities for the continuous spectrum Σ_α. From left to right, a linearly stable, linearly unstable and marginal spectrum.

In the last case, if $\sigma(\xi_0)$ is a point on the imaginary axis, that is, if

$$\operatorname{Re}\sigma(\xi_0) = 0 ,$$

we say that this point is *marginally stable* if

$$\partial_\xi \sigma(\xi_0) \neq 0 ,$$

and *marginally indifferent* otherwise. Recall that the condition of marginality means that in addition to the above conditions, the spectrum must lie in the closed left half-space. This means that

$$\partial_\xi \operatorname{Re}\sigma(\xi_0) = 0 ,$$

is a necessary condition for marginality. More precisely, the condition is

$$\begin{aligned} &\partial_\xi^k \operatorname{Re}\sigma(\xi_0) = 0 , \quad \text{for } k = 1, \ldots, 2m-1 , \\ &\partial_\xi^{2m} \operatorname{Re}\sigma(\xi_0) \neq 0 , \end{aligned} \tag{16.1}$$

for some $m > 0$ (assuming σ is sufficiently differentiable).

We have already explained in the example above why the definitions of stability and instability are adequate. We now explain the difference between the marginally stable and the marginally indifferent situation. The method of stationary phase will be used below to investigate the behavior of the solution for the marginal case. This is a well-known and difficult problem connected with catastrophe theory. We refer to the recent monograph of Arnold, Varchenko and Goussein-Zadé [AVG] for an exposition of the subject. However, in our case, we will only ask simple questions.

We consider the marginally stable situation, that is, we assume that

$$\mathrm{Re}\,\sigma(\xi_0) = 0 \quad \text{and} \quad \partial_\xi \sigma(\xi_0) \neq 0 \,,$$

and that this derivative depends continuously on ξ near ξ_0. In addition, we assume that (16.1) holds. Under these assumptions, we see that any initial perturbation v in the Schwartz space $\mathcal{S}$ of tempered functions has a linear time evolution given by

$$v(x,t) = \int_{-\infty}^{\infty} d\xi \, e^{t\sigma(\xi)} \, e^{i\xi x} v(\xi) \,. \tag{16.2}$$

(In the case of several components, we also require that $v(\xi)$ is proportional to an eigenvector of the matrix valued operator A, see below for a concrete example.) We now use the identity

$$\int_{-\infty}^{+\infty} d\xi \, e^{i\xi x} v(\xi) e^{t\sigma(\xi)} = \frac{1}{t} \int_{-\infty}^{+\infty} d\xi \, \frac{e^{i\xi x} v(\xi)}{\partial_\xi \sigma(\xi)} \frac{d}{d\xi} e^{t\sigma(\xi)} \,.$$

After an integration by parts we see that the above integral equals

$$-\frac{1}{t} \int_{-\infty}^{\infty} d\xi \frac{d}{d\xi} \left(\frac{e^{i\xi x} v(\xi)}{\partial_\xi \sigma(\xi)} \right) e^{t\sigma(\xi)} \,.$$

From the assumption $\partial_\xi \sigma(\xi_0) \neq 0$ it is clear that for functions v with support close enough to ξ_0 this integral decays at least like $1/t$ for large t. (For v with support in the complement of a neighborhood of ξ_0 – or any other such points – the spectrum of $A(\xi)$ lies strictly in the left half-plane and the integral is seen to decay exponentially with t.) Moreover, one can repeat the argument to show that the integral decreases faster than any power of t. This shows that in the marginally stable situation, a perturbation tends to zero if its Fourier transform is sufficiently differentiable. In direct space this means that sufficiently *localized* perturbations decay. The above argument breaks down if $\partial_\xi \sigma(\xi_0) = 0$, and convergence or divergence depends on details of v and the higher derivatives of σ. (If one assumes some analyticity properties of the Fourier transform of v, one can still show exponential decay, and depending on the relation between the domain of analyticity and the real part of $\sigma(\xi)$, convergence may even occur for positive (but small) values of $\operatorname{Re}\sigma(\xi)$; see Huerre [Hu] for a general discussion. This seems to be at first in contradiction with the definition of stability explained before; however, we insist that *any instability* is only visible in sufficiently large function spaces. On the other hand, if the function space is too big, everything is unstable, cf. [S].)

We emphasize that the above considerations are only valid for test functions $v(\xi)$ which are regular enough. In position space, this corresponds to considering localized perturbations. In particular, the above considerations do not apply when the perturbation $v(\xi)$ is a δ-function, that is, when its Fourier transform is periodic. We shall come back to this important point, and it will be seen that the apparent speed of propagation depends on the class of functions v one considers.

16.2. Moving Frames

As explained before, we shall use frames moving at a constant speed c with respect to the laboratory frame for which the evolution equation was first written. If $u(x,t)$ is a solution of the evolution equation in the laboratory frame, it becomes $u(x+ct,t)$ when it is observed from the frame moving at speed c. One can check that the evolution equation in the moving frame is given by

$$\partial_t u = \mathcal{N}(u) - c \cdot \partial_x u \ ,$$

where $\cdot$ denotes the scalar product of $\mathbf{R}^d$.

One can repeat the linear stability analysis in the moving frame. Changing to the moving frame, we see that the operator A is replaced by $A - c \cdot \partial_x$, or, in Fourier space, $A - ic \cdot \xi$. It is easy to verify that if the stationary solution is stable in the laboratory frame, it is also stable in any moving frame, because the eigenvalues of A and of $A - ic \cdot \xi$ differ only by the imaginary quantity $ic \cdot \xi$. Hence the real parts do not change sign, and the nature of the (in)stability remains unchanged.

The situation is different in the marginal case. We now show how in this case, *the frame of reference in which we observe decides on marginal stability or marginal indifference.* We see that in the moving frame the eigenvalue is $\sigma(\xi) - ic \cdot \xi$. Therefore, *depending on the frame of reference in which it is observed*, a marginal system looks either stable or indifferent.

Definition. A marginal system is called *convectively unstable* if $\partial_\xi \sigma(\xi_0) \neq 0$. In this case, $\partial_\xi \sigma(\xi_0)$ is purely imaginary and we define $c = -i\partial_\xi \sigma(\xi_0)$.

According to what we said before, if a convectively unstable system really develops an instability, then this instability will move with speed c in the laboratory frame. It is "convected away," and this phenomenon is responsible for the name of this instability.

It is natural to ask if convective instabilities are connected with bifurcations from the stationary solution. From what we have said above, it is tempting to look for bifurcated solutions moving at the speed of the convective instability. However, there is another argument which suggests that the speed of the bifurcated solution is equal to $c = \operatorname{Im} \sigma(\xi_0)$. The argument goes as follows: If we consider the linearized evolution equation in a frame moving at this speed c, then at the bifurcation value the function $e^{i\xi_0 x}$ is a stationary solution. We shall show later that this phase is indeed the dominant part of a bifurcated solution to the nonlinear problem. The reason for the difference in speeds is in fact easy to understand. The stationary phase method can only be applied to sufficiently regular test functions. Assume now that we want to consider a test function which is a phase in direct space: $v(x) = e^{iqx}$. This function is not in the space $\mathcal{S}$ of tempered functions. We have in Fourier space $v(\xi) = \delta(\xi - q)$. Therefore, using equation (16.2) we get

$$v(x, t) = e^{t\sigma(q)} e^{iqx} .$$

If $q \neq \xi_0$ this function decays exponentially fast in time. On the other hand, if $q = \xi_0$ it oscillates (in time) in the laboratory frame but it is fixed in the frame moving at speed $\operatorname{Im} \sigma(\xi_0)$. This is a clear sign of a bifurcation.

16.3. Some Examples

The real amplitude equation. We have already discussed part of this case above, but now we give a complete enumeration of the various possibilities. The equation is

$$\partial_t u(x, t) = \partial_x^2 u(x, t) + \alpha u(x, t) - u^3(x, t) . \tag{16.3}$$

The linearized equation near the solution $u \equiv 0$ is

$$\partial_t v(x,t) = (\alpha + \partial_x^2) v(x,t) , \tag{16.4}$$

so that the spectrum is the set

$$\Sigma = \{\alpha - \xi^2 : \xi \in \mathbf{R}\} .$$

When $\alpha < 0$, the system is linearly stable, when $\alpha > 0$, it is linearly unstable, and then all infinitesimal perturbations with Fourier modes satisfying $\xi^2 < \alpha$ will diverge exponentially with rate $\alpha - \xi^2$. When $\alpha = 0$ the system is marginally indifferent. In fact, the Fourier component of the eigenfunction is 0, and any constant function is fixed in time for the evolution equation (16.4). Under the nonlinear evolution (16.3), a small constant initial condition will grow to the point where $u \equiv \pm\alpha^{1/2}$. Note that the nature of the nonlinearity, for example its sign or degree, is irrelevant for the linear analysis, but, of course, not for the nonlinear evolution. We shall see later that *every* nonzero initial condition satisfying $0 \le u(x) \le \alpha^{1/2}$ will tend to $u \equiv \alpha^{1/2}$.

The SH equation. This equation is

$$\partial_t u(x,t) = \big(\alpha - (1 + \partial_x^2)^2\big) u(x,t) - u^3(x,t) .$$

The linearized equation near the solution $u \equiv 0$ is

$$\partial_t v(x,t) = \big(\alpha - (1 + \partial_x^2)^2\big) v(x,t) ,$$

so that the spectrum is the set

$$\Sigma = \{\sigma(\xi) = \alpha - (1 - \xi^2)^2 : \xi \in \mathbf{R}\} .$$

For $\alpha < 0$ the system is linearly stable and for $\alpha > 0$ it is linearly unstable. When $\alpha = 0$, then $\operatorname{Re}\sigma(\xi_0) = 0$ for $\xi_0 = \pm 1$. The derivative of $\sigma(\xi)$ satisfies $\partial_\xi \sigma(\xi_0) = 0$. Therefore, the system is marginally indifferent for speed $c = 0$. This means that it is not convectively unstable. The potentially unstable modes in the marginal situation are the functions $\cos(x)$ and $\sin(x)$. They are not solutions of the nonlinear problem, and will thus grow to a periodic stationary solution. In Section 17 we will construct this solution, but we shall not discuss the time-dependence.

A Convective Instability. We now describe a more subtle case. We set the space dimension to $d = 1$ and the order parameter dimension to $\nu = 2$. The evolution equation is given by

$$\partial_t u = \begin{pmatrix} 1 & 0 \\ 0 & 1 \end{pmatrix} \partial_x^6 u + \begin{pmatrix} 2 & 5 \\ -1 & 0 \end{pmatrix} \partial_x^3 u + \begin{pmatrix} \alpha - 1 & 0 \\ 0 & \alpha - 1 \end{pmatrix} u - \begin{pmatrix} u_1^3 \\ u_2^3 \end{pmatrix} ,$$

where u_1 and u_2 are the two components of the order parameter u.

The linearized evolution around the stationary solution $U = 0$ is given by

$$\partial_t v = \begin{pmatrix} 1 & 0 \\ 0 & 1 \end{pmatrix} \partial_x^6 v + \begin{pmatrix} 2 & 5 \\ -1 & 0 \end{pmatrix} \partial_x^3 v + \begin{pmatrix} \alpha - 1 & 0 \\ 0 & \alpha - 1 \end{pmatrix} v .$$

If we denote again by v the Fourier transform with respect to x of the function $v(x, t)$, we have

$$\partial_t v = Av \ ,$$

where the matrix valued function $A(\cdot)$ is given by

$$A(\xi) = \begin{pmatrix} -\xi^6 + 2i\xi^3 + \alpha - 1 & 5i\xi^3 \\ -i\xi^3 & -\xi^6 + \alpha - 1 \end{pmatrix} .$$

The unitary matrix

$$O = \frac{1}{\sqrt{6}} \begin{pmatrix} 2 - i & -2 - i \\ i & i \end{pmatrix}$$

diagonalizes $A(\xi)$ for all ξ. Namely

$$O^{-1} A(\xi) O = \begin{pmatrix} \sigma_+(\xi) & 0 \\ 0 & \sigma_-(\xi) \end{pmatrix} ,$$

where the two eigenvalues are given by

$$\sigma_\pm(\xi) = -(\xi^3 \mp 1)^2 + i\xi^3 + \alpha \ .$$

We observe that for $\alpha < 0$ both eigenvalues describe continuous curves contained in the open complex left half-plane when ξ varies over $\mathbf{R}$. For $\alpha = 0$, the two curves are tangent to the imaginary axis, and for $\alpha > 0$ they penetrate into the right half-plane. According to the above discussions we shall call $\alpha = 0$ a point of instability (bifurcation). We now discuss how this instability develops.

Let v be an initial perturbation belonging to the Schwartz space $\mathcal{S}$ of tempered functions. The perturbation at time t is given in Fourier variables by

$$v(x, t) = \int_{-\infty}^{+\infty} e^{i\xi x} O^{-1} \begin{pmatrix} e^{t\sigma_+(\xi)} & 0 \\ 0 & e^{t\sigma_-(\xi)} \end{pmatrix} Ov(\xi, 0)\, d\xi \ .$$

Each component of the vector $v(x,t)$ is given by a sum of two integrals of the form

$$\int_{-\infty}^{+\infty} e^{i\xi x} f(\xi) e^{t\sigma(\xi)} d\xi \ ,$$

where σ is equal to σ_+ or σ_-, and we want to know the asymptotic behavior of these integrals for large time.

First, if $\alpha < 0$, the integrals decrease exponentially in time because

$$\sup_{\xi \in \mathbf{R}} \operatorname{Re} \sigma_\pm(\xi) < 0 \ .$$

This is an indication that the stationary solution is stable. The case of $\alpha > 0$ similarly leads to an instability. We now come to the interesting case of $\alpha = 0$. The critical ξ are $\xi_0 = \pm 1$. Both lead to the same phenomena and therefore it suffices to pursue only the case $\xi_0 = 1$. Then

$$\begin{aligned} \operatorname{Re} \sigma_-(\xi_0) &= 0 \ , \\ \partial_\xi \sigma_-(\xi_0) &= 3i \ . \end{aligned}$$

Thus, we are in the case of a convective instability which will appear for a speed $c = 3$. In the frame moving with speed c, the linearized operator is

$$A(\xi) - ic\xi \mathbf{1} \ ,$$

where $\mathbf{1}$ is the identity matrix. It is still diagonalized by the matrix O but the eigenvalues are now $\sigma_\pm(\xi) - ic\xi$. If we consider again $\alpha = 0$, the integrals

$$\int_{-\infty}^{+\infty} f(\xi) e^{t(\sigma_\pm(\xi) - ic\xi)} d\xi$$

still decay faster than any power provided $c \neq 3$. However, for $c = 3$, the quantities $d\sigma_\pm(\xi)/d\xi$ are equal to zero for $\xi = \pm 1$ respectively. One can therefore find functions f in S for which the integral does not decay faster than some inverse power of t. The speed c of the convective instability at the threshold is always given by

$$c = \operatorname{Im} \partial_\xi \sigma(\xi_0) \ .$$

Note also that these instabilities correspond to well defined modes. For further comparison, it should be noted that in the present case, that is, for smooth test functions, the speed is given by the derivative $\operatorname{Im} \partial_\xi \sigma(\xi_0)$. Later, it will be seen that for the case of δ-functions, that is, for periodic solutions, it is related to $\operatorname{Im} \sigma(\xi_0)$. The example above has been carefully chosen to make the two speeds different.

17. Existence of Stationary Solutions

From the discussion of absolute and convective instability, we shall recall the three main examples. Our aim in this section is to discuss the actual bifurcation for these examples, and to illustrate to what extent the methods of perturbation theory and of the proofs are general.

Before we start, the reader should keep in mind that this is a less ambitious scheme than studying the *dynamical* (in)stability which was suggested in the analysis of the previous section. In fact, not much is known rigorously for this harder problem, but it still serves as the main motivation for the study of bifurcations and it determines, to a large extent, its terminology.

17.1. The Nonlinear Heat Equation

The first sample problem we study is

$$\partial_t u = \partial_x^2 u + \alpha u - u^3 \equiv \mathcal{N}_\alpha(u) . \tag{17.1}$$

This equation has a constant solution $U = 0$. We seek a *stationary* solution near $U = 0$, in the space of *time-independent* solutions of (17.1). In this trivial case, we see that a second solution bifurcates from the constant solution $U = 0$, namely, in the notation of Section 14.3, the bifurcating solution is given by

$$\alpha(\varepsilon) = \varepsilon^2 , \quad u(\varepsilon)(x) = \varepsilon ,$$

for all $\varepsilon \in \mathbf{R}$.

17.2. The SH Equation (Perturbation Theory and Statement of Results)

We consider next the more interesting, and nontrivial, case of the Swift-Hohenberg equation

$$\begin{aligned} \partial_t u(t,x) &= -(1+\partial_x^2)^2 u(t,x) + \alpha u(t,x) - u(t,x)^3 \\ &= -\partial_x^4 u(t,x) - 2\partial_x^2 u(t,x) + (\alpha - 1)u(t,x) - u(t,x)^3 \\ &\equiv \mathcal{N}_\alpha(u)(t,x) . \end{aligned} \tag{17.2}$$

For all α, the function $U \equiv 0$ is a stationary solution of (17.2). We will study bifurcations from this solution which are small amplitude **periodic solutions with period** $2\pi/\omega$. We begin by determining those parameter values α for which such a bifurcation can occur. We therefore consider, on the space of $2\pi/\omega$-periodic functions, the tangent map to $\mathcal{N}_\alpha$ (at $U \equiv 0$). This is the linear operator A_α given by

$$A_\alpha v(x) = -\partial_x^4 v(x) - 2\partial_x^2 v(x) + (\alpha - 1)v(x) .$$

On the space of $2\pi/\omega$-periodic functions, the spectrum is

$$\alpha - (1 - n^2\omega^2)^2 , \quad \text{for } n \in \mathbf{Z} . \tag{17.3}$$

Every value of α for which one of the above expressions vanishes is a potential bifurcation point. We shall assume from now on that $|\omega|$ is close to 1. Then, as α increases from 0, the *first* expression which will vanish is $\alpha - (1-\omega^2)^2$, that is, (17.3) for $n = \pm 1$. We shall therefore only study the bifurcation which is related to this case. We call

$$\mu(\omega) = (1 - \omega^2)^2 \tag{17.4}$$

the corresponding bifurcation value for α. Accordingly, we reparametrize α as

$$\alpha = \lambda + \mu(\omega) ,$$

so that λ measures the distance from the bifurcation point. Thus, λ will play the same role as in Section 14.3.

Instead of $\mathcal{N}_\alpha$ we consider now the operator

$$\mathcal{M}_{\omega,\lambda}(u) = (\lambda + \mu(\omega) - (1 + \partial_x^2)^2)u - u^3 .$$

The parameter ω will be fixed, and we restrict the operator to a space of $2\pi/\omega$-periodic functions. (It is customary to express the results in other variables, and we will do so later, but from the point of view of bifurcation theory, the above choice seems the most natural one.)

We note that, for all λ, we have

$$\mathcal{M}_{\omega,\lambda}(0) = 0 ,$$

that is, $u = 0$ solves $\mathcal{M}_{\omega,\lambda}(u) = 0$ for all λ. According to bifurcation theory, we have to consider the tangent map $A_{\omega,\lambda} = \partial_u \mathcal{M}_{\omega,\lambda}(0)$.

At this point, a more careful discussion of the function spaces is necessary. We also want to point out some of the important features of the operator $\mathcal{M}_{\omega,\lambda}$ which we use in our approach to the bifurcation problem.

S1. The first property concerns the linear operator $A_{\omega,\lambda}$. It is a *differential operator with constant coefficients*:

$$A_{\omega,\lambda} = P_{\omega,\lambda}(-i\partial_x) ,$$

with P a polynomial. In addition, the polynomial has the symmetry

$$P_{\omega,\lambda}(\xi) = \overline{P_{\omega,\lambda}(-\xi)} , \tag{17.5}$$

where $\bar{z}$ denotes the complex conjugate of z. In the example at hand, the polynomial has even more symmetries (it is real), but only the symmetry (17.5) will be used.

S2. The second property will imply that the bifurcated solutions are *periodic functions* of x. (We will partially relax this condition in later developments.) Informally stated, we require that the nonlinear operator leaves the space of periodic functions with period $2\pi/\omega$ invariant. In the example at hand this can be seen from the condition S1 for the linear part of the operator (a differential operator with constant coefficients maps periodic functions to periodic functions with the same period). For the nonlinear part, it follows from its simple form (the third power of a periodic function is periodic with the same period).

We denote by Ω'_ω the space of real, $2\pi/\omega$-periodic functions. To break the translation invariance, we also require that the functions have a vanishing Fourier component $\sin(\omega x)$. This defines a real-linear subspace Ω_ω of Ω'_ω Therefore, the general element of Ω_ω has a Fourier expansion of the form

$$\Omega_\omega = \left\{ v \ : \ v(x) = \sum_{n \in \mathbf{Z}} \eta_n e^{in\omega x} \right\} ,$$

with $\eta_{-n} = \bar{\eta}_n$, and $\eta_1 \in \mathbf{R}$. Later, Ω_ω will be equipped with a norm and made it into a (real) Banach space. We shall sometimes identify Ω_ω with the set of its Fourier coefficients.

After these precautions, we now verify that the operator $\mathcal{M}_{\omega,\lambda}$ satisfies the conditions of Theorem 14.1. First of all, it acts on the space $\mathbf{X} = \Omega_\omega$, and leaves it invariant (so far we have only checked the periodicity, below, we shall also check the norm). In the setting of Theorem 14.1, we have $\mathbf{Y} = \mathbf{X}$. By the equations (17.3) and (17.4), the operator $\partial_u \mathcal{M}_{\omega,\lambda}$ has the isolated, simple eigenvalue 0 for $\lambda = 0$. The corresponding eigenvector is $v_0 = 2\cos(\omega x)$. Here, we use that Ω_ω only contains $\cos(\omega x)$ but not $\sin(\omega x)$. Next, we consider the mixed derivative and we see that

$$\partial_\lambda \partial_u \mathcal{M}_{\omega,\lambda=0}(0) v_0 = v_0 ,$$

since the operator is the identity, and hence the transversality condition of Theorem 14.1 is satisfied. It follows that there is a one-parameter family of solutions of

$$\mathcal{M}_{\omega,\lambda(\varepsilon)}(u(\varepsilon)) = 0 ,$$

which are of the form

$$u(\varepsilon)(x) = \varepsilon v_0(x) + \varepsilon \eta(\varepsilon)(x) ,$$

with

$$\eta(\varepsilon)(x) = \sum_{n \geq 0, n \neq 1} \eta_n(\varepsilon) 2\cos(n\omega x) .$$

(In fact, in our concrete example, all terms with even n will be absent, because the nonlinearity is odd.)

Perturbation Theory. Before going into technical details, we describe how perturbation theory is started in the case of the Swift-Hohenberg equation. Having set $\mu(\omega) = (1-\omega^2)^2$, the equation is now

$$\big(\lambda + (1-\omega^2)^2 - (1+\partial_x^2)^2\big)u(x) - u^3(x) = 0 . \qquad (17.6)$$

We substitute $u(x) = 2\varepsilon\cos(\omega x)$. The nonlinearity is, with this ansatz,

$$u^3(x) = 6\varepsilon^3\cos(\omega x) + 2\varepsilon^3\cos(3\omega x) .$$

If we consider only the frequency ω, that is, if we just neglect the last term, then this leads to

$$\big(\lambda + (1-\omega^2)^2 - (1-\omega^2)^2\big)2\varepsilon\cos(\omega x) - 6\varepsilon^3\cos(\omega x) = 0 .$$

Therefore, to lowest order in ε, we find $\lambda = 3\varepsilon^2$. The next orders are now obtained by substituting this solution in (17.6) and iterating (without neglecting higher frequencies).

Theorem 17.1. *Fix $\rho > 1$, $d_2 > 0$ and $d_1 > 1$. Assume that ω satisfies*

$$2d_1/5 < \omega^2 < 2 - d_2 \, . \tag{17.7}$$

Set $\mu(\omega) = (1-\omega^2)^2$. Then, there is an $\varepsilon_0 > 0$ depending only on ρ, d_1, d_2, such that for all ε satisfying $|\varepsilon| < \varepsilon_0$, the equation

$$(\lambda + \mu(\omega) - (1+\partial_x^2)^2)u(x) - u^3(x) = 0 \tag{17.8}$$

has a unique solution of the form

$$\lambda = \lambda(\varepsilon) = 3\varepsilon^2 + \mathcal{O}(\varepsilon^3) \, , \tag{17.9}$$

$$u(x) = u(\varepsilon)(x) = \varepsilon\, 2\cos(\omega x) + \sum_{n\geq 0, n\neq 1} \eta_n(\varepsilon) 2\cos(\omega n x) \, , \tag{17.10}$$

with $|\eta_n| \leq \text{const.}\, \varepsilon^2 \rho^{-n}$.

Remark. One can see from the construction of the solution that the bound on η_n is really $|\eta_n| \leq \mathcal{O}(\rho^{-n}\varepsilon^{||n|-1|})$.

Remark. If $\rho > 1$, then the Fourier series for u has rapidly vanishing coefficients and thus describes a solution which is analytic in the domain $|\mathrm{Im}\, z| < (\log \rho)/\omega$. From the preceding remark it also follows that the domain of analyticity in fact goes to the whole complex plane as ε tends to zero.

Remark. By (17.9), we see that for small ε, the quantity $\lambda(\varepsilon)$ is *positive*. This means that the equation (17.8) has – for small amplitudes – solutions only for values of λ, and μ satisfying

$$\lambda + \mu - (1-\omega^2)^2 > 0 \, .$$

This, in turn, implies that *the equation*

$$\alpha u(x) - (1+\partial_x^2)^2 u(x) - u^3(x) = 0$$

has small amplitude solutions of frequency ω (in the direction $\cos(\omega x)$*), for ω satisfying (17.7) exactly when*

$$\alpha > (1-\omega^2)^2 \, .$$

The amplitudes of these solutions are described by

$$u(x) = 2\cdot 3^{-1/2}\sqrt{\alpha - (1-\omega^2)^2}\cos(\omega x) + \mathcal{O}(|\alpha - (1-\omega^2)^2|) \, . \tag{17.11}$$

17.3. The SH Equation (Existence Proof)

We define the space Ω_ω as the Banach space of periodic functions of the form

$$u(x) = \sum_{n\in\mathbf{Z}} u_n e^{in\omega x} ,$$

with $u_n = \bar{u}_{-n} \in \mathbf{C}$, and $u_1 \in \mathbf{R}$. We fix $\rho \geq 1$ and we equip Ω_ω with the norm

$$\|u\|_\rho = \sum_{n\in\mathbf{Z}} \rho^{|n|} |u_n| . \tag{17.12}$$

If u, v are in Ω_ω, and we denote $u \cdot v$ the function $u(x)v(x)$, then we have the bound

$$\begin{aligned} \|u \cdot v\|_\rho &= \sum_{n\in\mathbf{Z}} \rho^{|n|} \left| \sum_{p+q=n} u_p v_q \right| \leq \sum_{p,q\in\mathbf{Z}} \rho^{|p+q|} |u_p v_q| \\ &\leq \sum_{p,q\in\mathbf{Z}} \rho^{|p|} |u_p| \, \rho^{|q|} |u_q| = \|u\|_\rho \, \|v\|_\rho . \end{aligned} \tag{17.13}$$

The second inequality follows from $\rho \geq 1$ and $|p+q| \leq |p|+|q|$. Applying (17.13) twice, we see that $u \mapsto u^3$ is a bounded map from Ω_ω to itself. Since the derivatives of this operator are of a similar form, they are also bounded. We next study the operator

$$A = \partial_u \mathcal{M}_{\omega,\lambda}(0) = (\lambda + \mu(\omega))\mathbf{1} - (1 + \partial_x^2)^2 .$$

It is unbounded on Ω_ω, but its spectrum is discrete and is given by

$$\lambda + \mu(\omega) - (1 - n^2\omega^2)^2 , \quad n \in \mathbf{Z} .$$

Since we restrict to real functions, each eigenvalue is simple. Note that A does not satisfy the conditions of Theorem 14.1 because it is unbounded, but we shall follow the proof of Theorem 14.1 in spirit, while working directly with the contraction mapping principle. This method also shows the sense in which the assumptions of Theorem 14.1 are too strict. A more general formulation of Theorem 14.1 is thus possible, but we leave it to the reader. By our assumptions on $\mu(\omega)$ and ω, we see that for $\lambda = 0$, the operator A has the eigenvalue zero, with eigenvector $v_0 = 2\cos(\omega x)$ (the factor 2 will be convenient later). Recall that $\mathbf{X} = \mathbf{Y} = \Omega_\omega$. We choose as the supplementary space in $\mathbf{Y}$ the space

$$\mathbf{S} = \left\{ v \in \Omega_\omega \ : \ v(x) = \sum_{|n|\neq 1} v_n e^{in\omega x} \right\} ;$$

these are the Fourier components not already contained in v_0. Consider

$$\begin{aligned} u(x) &= \varepsilon 2\cos(\omega x) + \varepsilon \sum_{n\geq 0, n\neq 1} \eta_n 2\cos(n\omega x) \\ &= \varepsilon v_0(x) + \varepsilon\eta(x) \ . \end{aligned} \tag{17.14}$$

The equation we want to solve is

$$\left(\lambda + \mu(\omega) - (1+\partial_x^2)^2\right) u - u^3 = 0 \ , \tag{17.15}$$

in the unknowns λ and η. Although the function u is real, it is more convenient to write it as

$$u(x) = \varepsilon(e^{i\omega x} + e^{-i\omega x}) + \varepsilon \sum_{n\in \mathbf{Z},\, |n|\neq 1} \eta_n e^{i\omega n x} \ ,$$

with $\bar{\eta}_{-n} = \eta_n$, in fact, we will find $\eta_n \in \mathbf{R}$. To unify the notation, we define $\eta_1 = \eta_{-1} = 0$. Define furthermore the operators $G_n : \mathbf{S} \to \mathbf{C}$ by

$$G_n(\eta) = \left((v_0 + \eta)^3\right)_n \ , \tag{17.16}$$

where $(\cdot)_n$ denotes the coefficient of $\exp(i\omega n x)$. Comparing the coefficients of the various Fourier modes, the equation

$$(\lambda + \mu(\omega) - (1+\partial_x^2)^2)u = u^3 \tag{17.17}$$

with u as in (17.14) is equivalent to the system

$$\lambda = \lambda + \mu(\omega) - (1-\omega^2)^2 = \varepsilon^2 G_1(\eta) \ , \quad \text{for } |n| = 1 \ , \tag{17.18}$$

$$(\lambda + \mu(\omega) - (1-n^2\omega^2)^2)\eta_n = \varepsilon^2 G_n(\eta) \ , \quad \text{for } |n| \neq 1 \ . \tag{17.19}$$

The first equality in (17.18) follows from the definition (17.4) of $\mu(\omega)$. Since

$$v_0 = 1\cdot e^{i\omega x} + 1\cdot e^{-i\omega x} \ ,$$

we have

$$\|v_0\|_\rho \leq 2\rho \ .$$

Therefore, we get, by (17.16) and using linearity and polarization,

$$\|G(\eta)\| \leq \left(2\rho + \|\eta\|_\rho\right)^3 \ , \tag{17.20}$$

$$\|\partial_\eta G(\eta)\delta\eta\|_\rho \leq 3\left(2\rho + \|\eta\|_\rho\right)^2 \|\delta\eta\|_\rho \ , \tag{17.21}$$

$$\|G(0)\|_\rho = \|v_0^3\|_\rho \leq \left(2\rho\right)^3 \ . \tag{17.22}$$

We next attack the question of the inverse of $\lambda + \mu(\omega) - (1 + \partial_x^2)^2$. We restrict the range of ω as in the statement of Theorem 17.1: We choose two constants $d_2 > 0$, and $d_1 > 1$, such that

$$\frac{2}{5} < \frac{2}{5}d_1 < \omega^2 < 2 - d_2 < 2 . \qquad (17.23)$$

With these constants, we have the bound

$$\inf_{|n|\neq 1} |\mu(\omega) - (1 - n^2\omega^2)^2| \geq \min(2d_2d_1/5, 12d_1(d_1 - 1)/5) \equiv 2D > 0. \qquad (17.24)$$

Consider now the map $\mathcal{L} : \mathbf{R} \times \mathbf{S} \to \mathbf{R} \times \mathbf{S}$, defined by

$$\begin{aligned}(\lambda, \eta) &\mapsto \mathcal{L}(\lambda, \eta)\\ &= \big(\varepsilon^2 G_1(\eta), \{\frac{\varepsilon^2}{\lambda + \mu(\omega) - (1 - n^2\omega^2)^2} G_n(\eta)\}_{|n|\neq 1}\big)\\ &\equiv \big(\mathcal{L}^{(1)}(\lambda, \eta), \mathcal{L}^{(2)}(\lambda, \eta)\big) .\end{aligned}$$

In the first equality, we use the Fourier components to define $\mathcal{L}$. Then we can rewrite (17.18), (17.19) as the fixed point problem

$$(\lambda, \eta) = \mathcal{L}(\lambda, \eta) . \qquad (17.25)$$

By construction, we find

$$\mathcal{L}(0,0) = \big(\varepsilon^2 (v_0^3)_1, \{\frac{\varepsilon^2}{\mu(\omega) - (1 - 9\omega^2)^2} (v_0^3)_n\}_{|n|\neq 1}\big) .$$

We have used that $(v_0^3)_n = 0$ if $|n| \notin \{1, 3\}$. If we equip the space $\mathbf{R} \times \mathbf{S}$ of (λ, η) with the norm $|\lambda| + \|\eta\|_\rho$, then we see from (17.22) and the inequality

$$|\eta_n| \leq \rho^{-|n|} \|\eta\|_\rho ,$$

that

$$\begin{aligned}|\mathcal{L}^{(1)}(0,0)| &\leq \varepsilon^2 (2\rho)^3 \rho^{-1} ,\\ \|\mathcal{L}^{(2)}(0,0)\|_\rho &\leq \varepsilon^2 (2\rho)^3 / (2D) .\end{aligned}$$

Define now the set

$$\mathcal{C} = \{(\lambda, \eta) \in \mathbf{R} \times \mathbf{S} \ : \ |\lambda| < D, \ \|\eta\|_\rho < R\} .$$

Using (17.21), we see that for $(\lambda, \eta), (\lambda', \eta') \in \mathcal{C}$, we have

$$\begin{aligned} |\mathcal{L}^{(1)}(\lambda, \eta) - &\mathcal{L}^{(1)}(\lambda', \eta')| \\ &\leq \varepsilon^2 \int_0^1 d\alpha \, |\partial_\eta G_1((1-\alpha)\eta + \alpha\eta')| \, \|\eta - \eta'\|_\rho \\ &\leq \varepsilon^2 3(2\rho + R)^2 \rho^{-1} \cdot \|\eta - \eta'\|_\rho \, . \end{aligned} \tag{17.26}$$

Similarly, using (17.21) and (17.24) to bound the inverse of A on $\mathbf{S}$, and its derivative, we get

$$\begin{aligned} \|\mathcal{L}^{(2)}(\lambda, \eta) - &\mathcal{L}^{(2)}(\lambda, \eta')\|_\rho \\ &\leq \varepsilon^2 \int_0^1 d\alpha \, \|\partial_\eta G((1-\alpha)\eta + \alpha\eta')\|_\rho / (2D - D) \cdot \|\eta - \eta'\|_\rho \\ &\leq \varepsilon^2 3(2\rho + R)^2 / (2D - D) \cdot \|\eta - \eta'\|_\rho \, , \end{aligned} \tag{17.27}$$

and

$$\begin{aligned} \|\mathcal{L}^{(2)}(\lambda, \eta) - &\mathcal{L}^{(2)}(\lambda', \eta)\|_\rho \\ &\leq \varepsilon^2 \int_\lambda^{\lambda'} d\alpha \, \|\{\frac{G_n(\eta)}{(\lambda + \mu(\omega) - (1 - n^2\omega^2)^2)^2}\}_{|n| \neq 1}\|_\rho \\ &\leq \varepsilon^2 \frac{(2\rho + R)^3}{(2D - D)^2} |\lambda - \lambda'| \, . \end{aligned} \tag{17.28}$$

Combining (17.26)–(17.28), we see that for sufficiently small $|\varepsilon|$, $\mathcal{L}$ is a contraction, and the contraction rate is $\mathcal{O}(\varepsilon^2)$. Combining this with the fact that $\mathcal{L}(0,0)$ is at $\mathcal{O}(\varepsilon^2)$ from the center of $\mathcal{C}$, we see that for sufficiently small $|\varepsilon|$, the operator $\mathcal{L}$ maps $\mathcal{C}$ to itself and is a contraction. Therefore, the equation (17.25) has a unique fixed point in $\mathcal{C}$. This fixed point is a solution to the equation (17.15), which is of the form $u(x) = \varepsilon v_0(x) + \varepsilon\eta(x)$ with $\eta \in \mathbf{S}$. This completes the proof of Theorem 17.1.

We comment finally on the proofs of the remarks following the statement of Theorem 17.1.

- To show the bound on λ, use the identity $(v_0^3)_1 = 3$.
- It is easy to see in perturbation theory that the coefficients satisfy $|\eta_n(\varepsilon)| < \varepsilon^{||n|-1|}$.
- Replacing ρ in all the proofs above by $\tau\varepsilon^{\delta - 2/3}$, with $\delta > 0$ and R by $R_0\varepsilon^{\delta - 2/3}$, one can check that the argument goes through. Therefore, we see that
$$|\eta_n(\varepsilon)| = \mathcal{O}((\tau/\varepsilon^{2/3})^{-||n|-1|}) \, .$$
Some more work leads to the bound $|\eta_n| \leq \mathcal{O}(\rho^{-n}\varepsilon^{||n|-1|})$.

17.4. Convective Stationary Solutions

Before we generalize our study to a "generic" setting in Section 18, we wish to gain some experience by studying the convective example of Section 16.3. There, we analyzed the time-dependent behavior, but now we are interested in the existence of stationary solutions.

We consider the bifurcation problem in $\mathbf{R}^2$, derived from the time dependent equation

$$\begin{aligned}\partial_t U &= \begin{pmatrix}1 & 0\\ 0 & 1\end{pmatrix}\partial_x^6 U - \begin{pmatrix}2 & 5\\ -1 & 0\end{pmatrix}\partial_x^3 U\\ &\quad+\begin{pmatrix}\alpha-1 & 0\\ 0 & \alpha-1\end{pmatrix}U - \begin{pmatrix}U_1^3\\ U_2^3\end{pmatrix},\end{aligned} \tag{17.29}$$

where U_1 and U_2 denote the two components of U. The constant α is real. We are interested in *convective stationary solutions*, that is, in **periodic solutions with period** $2\pi/\omega$ **in a frame moving with speed** $c \neq 0$. Thus, we seek solutions U of the form $U(x,t) = u(x - ct)$. Then (17.29) is transformed to

$$\begin{aligned}0 &= \begin{pmatrix}1 & 0\\ 0 & 1\end{pmatrix}(\partial_x^6 + c\partial_x)u - \begin{pmatrix}2 & 5\\ -1 & 0\end{pmatrix}\partial_x^3 u\\ &\quad+\begin{pmatrix}\alpha-1 & 0\\ 0 & \alpha-1\end{pmatrix}u - \begin{pmatrix}u_1^3\\ u_2^3\end{pmatrix}.\end{aligned} \tag{17.30}$$

We view the r.h.s. of (17.30) as a nonlinear operator $\mathcal{N}_{\alpha,c}$, acting on a space of functions on $\mathbf{R}$ and taking values in $\mathbf{C}^2$. The function $u = 0$ is a solution of (17.30), for all α and c. The tangent map $A_{\alpha,c} \equiv \partial_u \mathcal{N}_{\alpha,c}$ (at $u \equiv 0$) is a *real differential operator with constant coefficients.* We have

$$A_{\alpha,c}v = \begin{pmatrix}\partial_x^6 - 2\partial_x^3 + c\partial_x + (\alpha-1) & -5\partial_x^3\\ \partial_x^3 & \partial_x^6 + c\partial_x + (\alpha-1)\end{pmatrix}v\,. \tag{17.31}$$

Its Fourier transform is the (matrix valued) function

$$\begin{aligned}\xi &\mapsto P_{\alpha,c}(\xi)\\ &= \begin{pmatrix}-\xi^6 + 2i\xi^3 + ic\xi + (\alpha-1) & 5i\xi^3\\ -i\xi^3 & -\xi^6 + ic\xi + (\alpha-1)\end{pmatrix}.\end{aligned}$$

Because $A_{\alpha,c}$ is a real operator for real α and c, this function has the symmetry $P_{\alpha,c}(\xi) = \bar{P}_{\alpha,c}(-\xi)$. The eigenvalues of $A_{\alpha,c}$ come in pairs

σ and $-\bar{\sigma}$, with conjugate eigenvectors. In our particular example, the eigenvalues are

$$\sigma_{\pm}(\xi) = \alpha - (\xi^3 \mp 1)^2 + i(\xi^3 + c\xi) ,$$

with the corresponding eigenvectors

$$e_{\pm} = v_{\pm} e^{i\xi x} = \begin{pmatrix} \pm 2i - 1 \\ 1 \end{pmatrix} e^{i\xi x} .$$

The discussion will now follow quite closely the argument of Section 17.3. We *fix* a frequency ω. Then the spectrum of $A_{\alpha,c}$ is the set

$$\sigma_{\pm}(n\omega) , \quad \text{with } n \in \mathbf{Z} .$$

Every value of α, c for which one of the $\sigma_{\pm}(n\omega)$ vanishes is a potential bifurcation point for the nonlinear problem. We shall assume that $|\omega|$ is close to 1. There is a symmetry between positive and negative ω and we continue our discussion for the case of $\omega \approx 1$ only. (The other choice leads to the same solutions.) Then the first bifurcations will occur for $n = \pm 1$, when $\alpha = (\omega^3 - 1)^2$, and $c = -\omega^3/\omega$. We therefore define

$$\mu(\omega) = (\omega^3 - 1)^2 ,$$
$$\gamma(\omega) = -\omega^3/\omega .$$

These are the bifurcation values for α and c. We correspondingly reparametrize α and c as

$$\alpha = \lambda + \mu(\omega) ,$$
$$c = \hat{c} + \gamma(\omega) .$$

Therefore, λ and $\hat{c}$ measure the distance of the real and imaginary parts from the bifurcation point.

Remark. The imaginary part of the spectrum is not "compensated" by a complex λ (resp. α), but by a change of the speed of the frame of reference. This is the origin of the convective nature of the instability.

In analogy with Section 17.3 we now define, instead of $\mathcal{N}_{\alpha,c}$ the operator $\mathcal{M}_{\omega,\lambda,\hat{c}}$ by

$$\mathcal{M}_{\omega,\lambda,\hat{c}}(u) = \left(\partial_x^6 + (\hat{c} + \gamma(\omega))\partial_x + \lambda + \mu(\omega) - 1\right) \begin{pmatrix} 1 & 0 \\ 0 & 1 \end{pmatrix} u - \begin{pmatrix} 2 & 5 \\ -1 & 0 \end{pmatrix} \partial_x^3 u - \begin{pmatrix} u_1^3 \\ u_2^3 \end{pmatrix} . \tag{17.32}$$

The parameter ω will be fixed, and we restrict the operator to a space of $2\pi/\omega$-periodic functions.

We note that, for all $\lambda, \hat{c}$, we have

$$\mathcal{M}_{\omega,\lambda,\hat{c}}(0) = 0 .$$

We consider therefore the tangent map

$$A_{\omega,\lambda,\hat{c}} = \partial_u \mathcal{M}_{\omega,\lambda,\hat{c}} ,$$

evaluated at $u \equiv 0$. Since we restrict to periodic functions with period $2\pi/\omega$, the operator $A_{\omega,\lambda,\hat{c}}$ has the double eigenvalue 0 for $\lambda = 0$, $\hat{c} = 0$. The corresponding complex eigenvectors are

$$\begin{pmatrix} 2i-1 \\ 1 \end{pmatrix} e^{i\omega x} \quad \text{and} \quad \begin{pmatrix} -2i-1 \\ 1 \end{pmatrix} e^{-i\omega x} .$$

One can combine them into a *real* eigenvector, given by

$$v_0 = 2 \begin{pmatrix} -2\sin(\omega x) - \cos(\omega x) \\ \cos(\omega x) \end{pmatrix} . \tag{17.33}$$

This eigenvalue is simple.

We next consider the derivatives $\partial_\lambda A_{\omega,0,0}$ and $\partial_{\hat{c}} A_{\omega,0,0}$. They are, respectively, the matrices $\mathbf{1}$ and $-i\omega\mathbf{1}$. Therefore, the derivative maps v_0 onto v_0. This means that the transversality condition of Theorem 14.2 is satisfied.

Remark. In this example, we are in a special situation because the two derivatives are proportional. We shall discuss in the next subsection an example where they are different. For the time being, we verify directly in perturbation theory that the transversality condition of Theorem 14.1 is satisfied. The difficulty in a proper treatment of these cases is connected to the study of real solutions. A general discussion must proceed with the complex subspace connected to the eigenvalue zero. It will be given in Section 27.4.

Assuming thus the transversality condition to be satisfied, we continue our analysis. We apply the methods of Section 17.3 to show that the 0 eigenvalue produces a bifurcation in the space of real functions u with period $2\pi/\omega$. This means that the equation (17.29) will have real solutions with "stationary amplitude" which are of the form

$$U(x,t) = \sum_{n\in\mathbf{Z}} \eta_n e^{i\omega n(x-ct)} ,$$

with $\eta_n = \bar{\eta}_{-n}$. We proceed in a way which is analogous to Section 17.3, but we adapt the basis to the eigenfunctions of $A_{\omega,0,0}$. We define the space $\mathbf{X} = \Omega_\omega$ as the set of functions $\mathbf{R} \to \mathbf{C}^2$ of the form

$$u(x) = \sum_{n \in \mathbf{Z}} u_n v_+ e^{i\omega nx} + \bar{u}_n v_- e^{-i\omega nx} ,$$

where the $v_\pm$ are again given by

$$v_+ = \begin{pmatrix} 2i - 1 \\ 1 \end{pmatrix} , \quad v_- = \begin{pmatrix} -2i - 1 \\ 1 \end{pmatrix} .$$

To break the translation invariance, we require $u_1 \in \mathbf{R}$. This restriction defines a subspace $\mathbf{X}'$ of $\mathbf{X}$. We further restrict $\mathbf{X}$ (and similarly $\mathbf{X}'$) by fixing $\rho \geq 1$ and requiring that for every $u \in \mathbf{X}$, the norm

$$\|u\|_\rho = \sum_{n \in \mathbf{Z}} \rho^{|n|} |u_n|$$

is finite. In the basis we have chosen, the eigenvector v_0 is given by $u_1 = 1$, and all other $u_n = 0$. We next define a space $\mathbf{S}$ which is supplementary in $\mathbf{X}'$ to the space spanned by the eigenvector v_0. It will be given by

$$\mathbf{S} = \{\eta \in \mathbf{X} : \eta_1 = 0\} . \tag{17.34}$$

As in the case of the Swift-Hohenberg equation, we consider separately the linear and the nonlinear parts of the problem. The map $U \mapsto (U_1^3, U_2^3)$ leads to the definition of an operator $G : \mathbf{S} \to \mathbf{X}$ given by

$$G(\eta) = \big((v_0 + \eta)_1^3, (v_0 + \eta)_2^3\big) . \tag{17.35}$$

The indices 1 and 2 refer to the two components of the vector in $\mathbf{R}^2$. We further define the restrictions G_n by the decomposition

$$G(\eta)(x) = \sum_{n \in \mathbf{Z}} G_n(\eta) v_+ e^{i\omega nx} + \bar{G}_n(\eta) v_- e^{-i\omega nx} .$$

This decomposition is unique since the $v_\pm \exp(\pm i\omega nx)$ form a basis of $\mathbf{X}$. The linear operator $A_{\omega,\lambda,\hat{c}}$ satisfies

$$\begin{aligned} A_{\omega,\lambda,\hat{c}} v_+ e^{i\omega nx} &= \big(\lambda + i\omega n\hat{c} + Q_+(\omega n)\big) v_+ \, e^{i\omega nx} , \\ A_{\omega,\lambda,\hat{c}} v_- e^{-i\omega nx} &= \big(\lambda - i\omega n\hat{c} + Q_-(-\omega n)\big) v_- \, e^{-i\omega nx} . \end{aligned}$$

where

$$Q_{\pm}(\xi) = \mu(\omega) + i\gamma(\omega)\xi - (\xi^3 \mp 1)^2 + i\xi^3 .$$

Note that by construction, $Q_{+}(\omega) + \mu(\omega) + i\gamma(\omega)\omega = Q_{-}(-\omega) + \mu(\omega) + i\gamma(\omega)\omega = 0$, and that for all other combinations of the sign s and the harmonics n, the quantity $Q_s(n\omega)+\mu(\omega)+i\gamma(\omega)n\omega$ is uniformly bounded away from 0 (for fixed ω, if $|\omega|$ is not too far from 1). The equation (17.30), with u of the form

$$u(x) = \varepsilon v_0 + \varepsilon\eta(x) ,$$

and with $\eta \in \mathbf{S}$, can then be rewritten as the system

$$\lambda + i\hat{c}\omega = \lambda + i\hat{c}\omega + \mu(\omega) + i\gamma(\omega)\omega + Q_{+}(\omega) = \varepsilon^2 G_1(\eta) ,$$

for $n = 1$,

$$\big(\lambda + i\hat{c}n\omega + \mu(\omega) + i\gamma(\omega)n\omega + Q_{+}(\omega n)\big)\eta_n = \varepsilon^2 G_n(\eta) , \qquad (17.36)$$

for $n \neq 1$, $n \in \mathbf{Z}$. Note that the function $G(\eta)(\ .\ )$, viewed as a function of x, is not necessarily real if η is not real. In view of this, we expect to see, as a function of ε, a change in the value of λ (λ is real!), and of the (real!) speed $\hat{c}$. Below, we calculate these corrections to leading order. To prove the existence of solutions to equation (17.36), we proceed in analogy with Section 17.3. Set $\tau = \lambda + i\hat{c}\omega$, $\tau \in \mathbf{C}$, λ, $\hat{c}$, $\omega \in \mathbf{R}$. We define the operator $\mathcal{L}$ by

$$\begin{aligned}
&(\tau,\eta) \mapsto \mathcal{L}(\tau,\eta) \\
&\quad = \Big(\varepsilon^2 G_1(\eta), \{\frac{\varepsilon^2 G_n(\eta)}{\mathrm{Re}\,\tau + in\mathrm{Im}\,\tau + \mu(\omega) + in\gamma(\omega)\omega + Q_{+}(\omega n)}\}_{|n|\neq 1}\Big) \\
&\quad \equiv \big(\mathcal{L}^{(1)}(\tau,\eta),\ \mathcal{L}^{(2)}(\tau,\eta)\big) .
\end{aligned}$$

It is an easy exercise to adapt the calculations of Section 17.3 to the present case. The operator $\mathcal{L}$ is again a contraction of a suitable ball and hence has a fixed point. This fixed point is a solution of equation (17.36). This solution is unique in the ball.

Remark. The system of equations (17.36) seems, at first sight, overdetermined because we have to equate not only the coefficients of $v_{+}e^{in\omega x}$ but also those of $v_{-}e^{-in\omega x}$. Note, however, that for any $n \neq 1$, one has

$$\begin{aligned}
\big(\lambda + i\hat{c}n\omega + \mu(\omega) + i\gamma(\omega)n\omega + Q_{+}(\omega n)\big)\eta_n &= \varepsilon^2 G_n(\eta) , \\
\big(\lambda - i\hat{c}n\omega + \mu(\omega) - i\gamma(\omega)n\omega + Q_{-}(-\omega n)\big)\bar{\eta}_n &= \varepsilon^2 \bar{G}_n(\eta) ,
\end{aligned}$$

The second equation is the complex conjugate of the first. A similar pair of equations occurs for $n = 1$. It should be noted that $n = -1$ is not a resonating case: The sum

$$\lambda - i\hat{c}\omega + \mu(\omega) - i\gamma(\omega)\omega + Q_+(-\omega)$$

is bounded away from zero, when ω is sufficiently close to 1.

It is perhaps instructive to compute the leading terms of the solution. For this, we need to compute the inhomogeneity $G(0) = \big((v_0)_1^3, (v_0)_2^3\big)$. For the determination of λ and $\hat{c}$ to lowest order in perturbation theory, we need only to know the component G_1. The coefficient of frequency ω in $G(0)$ is

$$3\begin{pmatrix}(2i-1)^2(-2i-1)\\ 1\end{pmatrix} e^{i\omega x} + 3\begin{pmatrix}(-2i-1)^2(2i-1)\\ 1\end{pmatrix} e^{-i\omega x}\ . \tag{17.37}$$

Redecomposing this along the basis $v_\pm$ in the form

$$\begin{aligned} &s v_+ e^{i\omega x} + \bar{s} v_- e^{-i\omega x} \\ &+ t v_+ e^{-i\omega x} + \bar{t} v_- e^{i\omega x}\ , \end{aligned}$$

we find $s = 9 + 3i$, $t = -6 + 3i$. The equation for λ and $\hat{c}$ is, to lowest order in ε,

$$\lambda + i\omega\hat{c} = \varepsilon^2 s\ .$$

Therefore, this leads to

$$\begin{aligned} \lambda(\varepsilon) &= 9\varepsilon^2 + \mathcal{O}(\varepsilon^3)\ , \\ \hat{c}(\varepsilon) &= \frac{3\varepsilon^2}{\omega} + \mathcal{O}(\varepsilon^3)\ . \end{aligned}$$

Expressed in the original variables, we see that the equation (17.30) has a solution of the form

$$2\varepsilon\begin{pmatrix}-2\sin(\omega x) - \cos(\omega x)\\ \cos(\omega x)\end{pmatrix} + \mathcal{O}(\varepsilon^2)\ ,$$

when

$$\begin{aligned} \alpha &= \mu(\omega) + \lambda = (\omega^3 - 1)^2 + 9\varepsilon^2 + \mathcal{O}(\varepsilon^3)\ , \\ c &= \gamma(\omega) + \hat{c} = -\omega^3/\omega + 3\varepsilon^2/\omega + \mathcal{O}(\varepsilon^3)\ . \end{aligned}$$

17.5. Convective Bifurcation (Second Example)

In Section 17.4 we encountered an equation with a slight degeneracy, namely, the tangent map was the same for the real and the imaginary part. We now present an example where this relation does not hold. To keep the computation of eigenvalues sufficiently simple, we choose a somewhat contorted parametrization. Fix ω near 1 and let

$$\mu(\omega) = (\omega^3 - 1)^2 , \quad \gamma(\omega) = -\omega^2 ,$$

as before. We now consider the nonlinear equation $\mathcal{N}_{\omega,\lambda,\hat{c}} = 0$, where

$$\begin{aligned} &\mathcal{N}_{\omega,\lambda,\hat{c}}(u) \\ &\quad = \begin{pmatrix} 1 & 0 \\ 0 & 1 \end{pmatrix} (\partial_x^6 + (\hat{c} + \gamma(\omega))\partial_x) u \\ &\quad + \begin{pmatrix} \lambda + \mu(\omega) - 1 & 0 \\ 0 & \mathbf{2}\lambda + \mu(\omega) - 1 \end{pmatrix} u \\ &\quad - \begin{pmatrix} 2 & 5 \\ -1 & 0 \end{pmatrix} \partial_x^3 u - \begin{pmatrix} u_1^3 \\ u_2^3 \end{pmatrix} . \end{aligned} \tag{17.38}$$

The only new feature with respect to (17.30) is the factor 2 (printed in bold type). We define the tangent map $\mathcal{A}_{\omega,\lambda,\hat{c}} = \partial_u \mathcal{N}_{\omega,\lambda\hat{c}}$, at $u \equiv 0$. Note that by our choice of (17.38) the operator $A_{\omega,0,0}$ has the *same* eigenvalues and eigenvectors as in Section 17.4. However, the derivatives with respect to the parameters are now different:

$$\begin{aligned} \partial_\lambda A_{\omega,0,0} &= \begin{pmatrix} 1 & 0 \\ 0 & 2 \end{pmatrix} , \\ \partial_{\hat{c}} A_{\omega,0,0} &= i\omega \begin{pmatrix} 1 & 0 \\ 0 & 1 \end{pmatrix} . \end{aligned}$$

For full generality, we should really consider an example with two non-commuting matrices. However, this would not introduce any new complications. We now study the solution of the problem for the case $n = \pm 1$. This leads to the *two complex equations* :

$$\begin{aligned} (\partial_\lambda A_{\omega,0,0} + \partial_{\hat{c}} A_{\omega,0,0}) v_+ &= \varepsilon^2 P_+(v_0^3) , \\ (\partial_\lambda A_{\omega,0,0} + \partial_{\hat{c}} A_{\omega,0,0}) v_- &= \varepsilon^2 P_-(v_0^3) , \end{aligned}$$

where $P_\pm$ is the projection onto $v_\pm$, and v_0 is defined as in (17.33). An explicit calculation then leads to a *unique* solution for the real unknowns λ and $\hat{c}$.

18. The General Case

We now consider a somewhat more general situation by analyzing what is really used in the proof of Theorem 17.1. Consider the problem

$$\mathcal{N}(\alpha, u) = 0 ,$$

and assume that

$$\mathcal{N}(\alpha, 0) = 0 ,$$

for all α near zero and

$$\partial_u \mathcal{N}(\alpha, 0) = L(\alpha) ,$$

where $L(\alpha)$ is a differential operator with constant coefficients. We assume that $L(\alpha)$ has the form $L(\alpha) = P(\alpha, -i\partial_x)$, where the matrix valued polynomial $P(\alpha, \cdot)$ satisfies $P(\alpha, \xi) = \overline{P(\alpha, -\xi)}$. But we do not assume that P is real, nor that u (and hence $\mathcal{N}$, and P) take values in $\mathbf{R}$: They may now take values in $\mathbf{R}^\nu$. We assume furthermore that for some α_0 and for some $\xi_0 \neq 0$ one has

$$0 \in \operatorname{Re} \Sigma\big(P(\alpha_0, \xi_0)\big) ,$$

but that 0 is not an eigenvalue, that is, the matrix $P(\alpha_0, \xi_0)$ is invertible. Recall that Σ denotes the spectrum. We further assume that a purely imaginary eigenvalue only occurs for $\xi = \pm\xi_0$, that when it occurs it is simple, and that for $\alpha < \alpha_0$, and for any ξ, $\operatorname{Re} \Sigma\big(P(\alpha, \xi)\big) < 0$. We also assume a transversality condition on the matrix $\partial_\alpha P(\alpha_0, \xi_0)$, namely that if we denote by v_+ a unit vector in the one dimensional kernel of $P(\alpha_0, \xi_0)$, then the spectral projection of the vector $\partial_\alpha P(\alpha_0, \xi_0) v_+$ on the spectral subspace corresponding to the purely imaginary eigenvalue is not a real multiple of iv_+.

We shall use the vector $v_- = \overline{v_+}$ which is a unit vector in the one dimensional kernel of $P(\alpha_0, -\xi_0)$. We expect to find a bifurcation in a direction belonging to the kernel of the operator $L(\alpha_0) + c_0\partial_x$ for a suitably chosen speed c_0. However, this kernel is a two dimensional real space. Indeed, it consists of vectors of the form

$$\rho\big((v_+ + v_-)\cos(\omega x + \phi) + i(v_+ - v_-)\sin(\omega x + \phi)\big) ,$$

where ρ, $\phi \in \mathbf{R}$. Note that the parameter ϕ reflects the translation invariance of the original problem.

In order to eliminate this degeneracy, we shall impose a supplementary equation for the problem. This can be viewed as a sort of gauge fixing constraint. Once we have proven the existence of a bifurcated solution for this special constraint, it is easy to use the translation invariance of the equation to get other (translated) solutions. There are, of course, many equivalent ways of fixing the translation degeneracy, we shall choose one which is particularly well suited to the application of Theorem 14.2.

We shall first simplify somewhat the notation by translating the parameters of the problem. By continuity, for any ω near ξ_0, there is a smallest value of the parameter $\alpha = \mu(\omega)$ such that

$$0 \in \operatorname{Re} \Sigma\big(P(\mu(\omega), \omega)\big) ,$$

and the transversality condition is still satisfied. We shall reparametrize α by setting $\alpha = \mu(\omega) + \lambda$. Similarly, we shall denote by $i\omega\gamma(\omega)$ the purely imaginary eigenvalue of $P(\mu(\omega), \omega)$. The speed c of the bifurcated solution will be more conveniently parametrized by $c = \gamma(\omega) + \hat{c}$.

Assume next that $\mathcal{N}$ does not depend explicitly on x. Then we can work again in the spaces Ω_ω of real periodic functions with period $2\pi/\omega$ (with values in $\mathbf{R}^\nu$), where ω is a value close to ξ_0.

We consider a new map $\mathcal{T}$ which is an extension of $\mathcal{N}$, and we shall verify for this map the hypotheses of Theorem 14.2. We define $\mathcal{T}$ as a map from a neighborhood $\mathcal{V}$ of the origin in $\mathbf{R}^2 \times \Omega_\omega$ to Ω_ω by

$$\mathcal{T}(\lambda, \hat{c}, u) = \mathcal{N}(\mu(\lambda + \omega), u) + (\hat{c} + \gamma(\omega))\partial_x u .$$

We now verify the hypotheses of Theorem 14.2. First, $u = 0$ is always a solution of $\mathcal{T}(\lambda, \hat{c}, u) = 0$. For $\lambda = \hat{c} = 0$, the differential $\partial_u \mathcal{T}(0,0,0)$ has a two dimensional kernel (generated by the vectors $(v_+ + v_-)\cos(\omega x)$ and $i(v_+ - v_-)\sin(\omega x)$). The range of the tangent map has codimension two in Ω_ω because it has codimension two in the Fourier sector $n = 1$ and codimension zero in the other sectors. We finally have to check condition d) of Theorem 14.2. Again, this is a condition which involves only the sector $n = 1$. We consider any nonzero real vector v_0 which is a linear combination of

$$v_1 = (v_+ + v_-)\cos(\omega x) + i(v_+ - v_-)\sin(\omega x)$$

and

$$v_2 = (v_+ + v_-)\sin(\omega x) - i(v_+ - v_-)\cos(\omega x) .$$

We have to verify that the two vectors $\partial_\lambda L(\mu(\omega))v_0$ and $\partial_x v_0$ generate a supplementary subspace of the range of $L(\mu(\omega))$ in the first Fourier sector. Note first that we have $\partial_x v_1 = -\omega v_2$ and $\partial_x v_2 = \omega v_1$. From the transversality assumption it follows that

$$\partial_\lambda L(\mu(\omega))v_1 = Rv_1 + Sv_2 + r_1 ,$$

and

$$\partial_\lambda L(\mu(\omega))v_2 = -Sv_1 + Rv_2 + r_2 ,$$

where R and S are real numbers and r_1 and r_2 are two vectors which are in the spectral subspace of $L(\mu(\omega))$ corresponding to the eigenvalues inside the open left half-plane for the first Fourier sector. The transversality assumption implies that $R \neq 0$, and condition d) of Theorem 14.2 is now easy to verify. Again, we are in a situation where the linear part of the operator is unbounded, but the same methods as in Section 17.3 show that the Theorem 14.2 generalizes to the case at hand.

CHAPTER V

CONSEQUENCES OF THE LINEAR INSTABILITY OF STATIONARY SOLUTIONS

The next few sections deal with a very general outlook on the stability analysis of stationary (and by analogy of convective) solutions. We pursue here the following ideas.

a) A linear stability analysis can be done in great generality, using the method of Bloch functions, or time $2\pi/\omega$ maps, when the stationary solution is periodic.

b) Because of invariances of the problem, the point 0 is seen to be in the spectrum of these excitations.

c) Depending on the bifurcation parameter ε and the spatial frequency ω, the spectrum near 0 may be stable or unstable. In the case of the SH equation, this instability is called the **Eckhaus instability**.

d) We show that when the system is *not* Eckhaus unstable it will have solutions which are exponentially decaying in space. They can be used to construct new stationary solutions for **inhomogeneous problems**. The dichotomy between c) and d) can be viewed as follows. *Below* the Eckhaus instability, excitations are of *exponential form* with a *real* exponent, and are thus *unbounded* in the domain $[-\infty, \infty]$, but bounded either on $[0, \infty]$ or on $[-\infty, 0]$. *Above* the instability the exponent becomes imaginary leading to *bounded* excitations, and hence the stationary solution is unstable with respect to *bounded* perturbations.

The following analysis shows how this phenomenon is related to the motion of eigenvalues in the complex domain as physical parameters are changed.

19. Linearized Equations

In this section, we study some general properties of linearized equations. We consider a semilinear equation

$$\partial_t U(x,t) = P(-i\partial_x)U(x,t) - f(U(x,t)) \,, \tag{19.1}$$

and we ask how "infinitesimal perturbations" of a stationary solution $U(x,t) = u(x)$ behave. The words infinitesimal perturbation mean the *linearized equation* obtained from (19.1) by substituting $U(x,t) = u(x)+ v(x,t)$, where u is as above, and expanding to first order in v. That is, we study the equation

$$\partial_t v(x,t) = P(-i\partial_x)v(x,t) - f'(u(x))v(x,t) \,. \tag{19.2}$$

The study of such equations will be important in two contexts:

- The stability analysis of quasistationary solutions.
- The asymptotic analysis of bifurcating solutions in half-spaces.

These problems will be studied in the next sections. Here, we concentrate on those properties which are independent of the detailed nature of either of the two problems. To be more specific, all of the analysis will be done for the Swift-Hohenberg equation. However, many features carry over without change to any semilinear equation. We proceed as follows, using the method of Bloch waves: We first study the operators $A_\kappa = e^{-i\kappa x}Ae^{i\kappa x}$, where A is the linearized evolution operator. As a function of physical parameters, the spectrum of A_κ changes. For the study of periodic solutions, the zero eigenvalues of A_κ will be important. For the study of bifurcation in half-spaces, the whole spectrum of A_κ is needed.

19.1. Perturbing Stationary Solutions of the SH Equation

We consider a periodic, time-independent solution $U(x,t) = u(x)$ of the Swift-Hohenberg equation

$$\partial_t U(x,t) = \alpha U(x,t) - (1+\partial_x^2)^2 U(x,t) - U(x,t)^3 \,, \tag{19.3}$$

which we constructed in Section 17.2. We fix α and consider all frequencies ω satisfying $\alpha > (1-\omega^2)^2$. According to perturbation theory, the solution is then of the form $u(x) \approx 2\varepsilon\cos(\omega x)$, where $\alpha - (1-\omega^2)^2 \approx 3\varepsilon^2$.

We now consider infinitesimal perturbations around this solution. The words "infinitesimal perturbation" mean the *linearized equation* obtained from (19.3) by substituting $U(x,t) = u(x) + v(x,t)$, where u is as above, and expanding to first order in v. Then one obtains an equation of the form

$$\partial_t v(x,t) = \alpha v(x,t) - (1+\partial_x^2)^2 v(x,t) - 3u(x)^2 v(x,t) \ . \qquad (19.4)$$

We begin by noting several symmetries of this equation.

First of all, the equation (19.3) does not depend explicitly on x. Therefore, the set of its solutions is invariant under translation, and $u(x+y)$ is, for every y, a solution of (19.3) (viewed as a function of x). Differentiating with respect to x, we see that $v(x,t) = \partial_x u(x)$ is a solution of (19.4).

Second, the equation (19.3) only contains even derivatives in x. Therefore, it is invariant under the change of coordinates $x \to -x$.

We are interested in solutions of (19.4) whose asymptotic behavior is prescribed. There are two, related, ways to study such problems. In the first, one studies the generator of the time evolution on a set of *periodic* functions, multiplied with an exponential. This is the theory of Bloch waves. One studies the operator on the r.h.s. of (19.4) on functions which are of the form $v(x) = w(x)e^{i\kappa x}$, where w is a $2\pi/\omega$-periodic function and $\kappa \in [-\omega/\pi, \omega/\pi]$. In other words, we study the spectrum of the operator A_κ, given by

$$A_\kappa w(x) = \alpha w(x) - (1-(\kappa - i\partial_x)^2)^2 w(x) - 3u(x)^2 w(x) \ ,$$

and acting on $w \in \mathrm{L}^2_{\mathrm{per}}([0, 2\pi/\omega], dx)$. This setting is well adapted to Fourier analysis. Note that

$$A_0 e^{i\kappa x} w(x) = e^{i\kappa x} A_\kappa w(x) \ .$$

A second method is obtained by considering the transfer matrix. Let X denote an element in $\mathbf{C}^4$. One considers the equation

$$\partial_x X(x) = \mathbf{M}(x) X(x) \ , \qquad (19.5)$$

for the unknown vector valued function $X(\cdot)$, where

$$\mathbf{M}(x) = \begin{pmatrix} 0 & 1 & 0 & 0 \\ 0 & 0 & 1 & 0 \\ 0 & 0 & 0 & 1 \\ 0 & -2 & 0 & \alpha - 1 - 3u^2(x) \end{pmatrix} . \qquad (19.6)$$

Denote next by $X(v,x)$ the vector whose components are the derivatives

$$X(v,x) = \left(v(x), \partial_x v(x), \partial_x^2 v(x), \partial_x^3 v(x)\right) .$$

Then the solution, $X(v,x)$, of the equation (19.5) with initial condition $X(v,0)$ coincides with that obtained by integrating

$$0 = \alpha v(x) - (1+\partial_x^2)^2 v(x) - 3u(x)^2 v(x) \tag{19.7}$$

from $x=0$ to $x=x$ with the same initial conditions. We have simply replaced the fourth order differential equation by a system of 4 first order differential equations. Note now that the matrix $\mathbf{M}(x)$ is a periodic function in x, of period $2\pi/\omega$. (In fact, the period is π/ω since u^2 has half the period of u.) We now define a new matrix, $\mathbf{M}$, by

$$\mathbf{M}X(0) = X(2\pi/\omega) . \tag{19.8}$$

In other words, one integrates (19.5) from $x=0$ to $x=2\pi/\omega$ with initial data $X(0)=X$. Thus, $\mathbf{M}$ is the "time" $2\pi/\omega$ map, where x plays the role of time in the dynamical system whose (time-dependent) vector field is given by $\mathbf{M}(x)$.

In this context, the symmetries of the problem show up as follows: The translation invariance implies that $X(\partial_x u, 0)$ is an eigenvector with eigenvalue 0 of $\mathbf{M}$. The space reflection symmetry of (19.3) implies that if $X=(X_0,X_1,X_2,X_3)$ is an eigenvector of $\mathbf{M}$ with eigenvalue e^{σ}, then $X=(X_0,-X_1,X_2,-X_3)$ is an eigenvector with eigenvalue $e^{-\sigma}$. The relation between A_κ and $\mathbf{M}$ is as follows: If $A_\kappa v=0$ then $A_{\kappa=0}e^{i\kappa x}v=0$, so that

$$\begin{aligned}\mathbf{M}X(e^{i\kappa x}v(x),0) &= X(e^{i\kappa x}v(x), 2\pi/\omega)\\ &= X(e^{2\pi i\kappa/\omega}e^{i\kappa x}v(x),0)\\ &= e^{2\pi i\kappa/\omega}X(e^{i\kappa x}v(x),0) .\end{aligned}$$

Therefore, if A_κ has an eigenvalue 0 then $\mathbf{M}$ has an eigenvalue e^{σ} with $\sigma = 2\pi i\kappa/\omega$.

Recall now that u is an even function of x; this is how we broke the translation invariance in the solution of the bifurcation problem. Therefore, the eigenvector with eigenvalue zero is of the form

$$X(\partial_x u, 0) = (0,a,0,b) ,$$

with $a=\partial_x^2 u(0) \approx -2\varepsilon^2\omega^2$, and $b=\partial_x^4 u(0) \approx 2\varepsilon^2\omega^4$. We shall see later that the symmetry leads to a double eigenvalue, but with a single eigenvector.

19.2. Perturbation Theory

In the developments of later sections, we need information on the spectrum of the operators A_κ. This will be done in detail later. However, it is extremely useful to have a qualitatively correct idea of how the operators A_κ act. The best method to gain this insight is perturbation theory. We compute the operator A_κ as a function of ε, κ, and ω, and we will, throughout this subsection, neglect higher order terms in ε.

To study the spectrum of A_κ, one considers the equation

$$A_\kappa w - \sigma w = f \,, \tag{19.9}$$

and one asks for which values of σ this equation has a unique solution. Those values of σ are then *not in the spectrum of* A_κ. We begin by decomposing w and f into their Fourier components,

$$w(x) = \sum_{n\in\mathbf{Z}} w_n e^{i\omega n x} \,.$$

Since we study real functions, we have $w_n = \bar{w}_{-n}$. The Eq. (19.9) is then equivalent to the system of equations, defined for $n \in \mathbf{Z}$,

$$\alpha w_n - (1-(\kappa+\omega n)^2)^2 w_n - \sigma w_n - 3\sum_{p+q=0} u_p u_q w_n$$
$$= 3\sum_{\substack{p+q\neq 0\\ p+q+r=n}} u_p u_q w_r + f_n \,. \tag{19.10}$$

First, we describe in more detail the "potential" which comes from u. Since $u_p = 0$ for even p, the system Eq. (19.10) decouples over the even and odd subspaces in n. It is useful to introduce the notation

$$T_m = 3\sum_{p+q=m} u_p u_q \,. \tag{19.11}$$

Note that $T_m = T_{-m}$, and that $T_m = 0$ when m is odd. From bifurcation theory, we know that u has an amplitude of order ε, where we now parametrize the solutions by

$$3\varepsilon^2 = \alpha - (1-\omega^2)^2 \,.$$

Here, we assume that α is small, but in the bifurcation theory of Section 17.3 we assumed that ε is small. Note, however, that because in the bifurcation region one has $\alpha > (1-\omega^2)^2$, the smallness of α implies that ε is small. We have also seen that u is approximately equal to

$$u(x) \approx 2\varepsilon \cos(\omega x) ,$$

cf. (17.11), and that all other Fourier modes of u are of order ε^2. In terms of T_m, this yields,

$$T_0 \approx 6\varepsilon^2 \tag{19.12}$$

and

$$T_2 = T_{-2} \approx 3\varepsilon^2 .$$

All other T_j are at least of order $\mathcal{O}(\varepsilon^3)$. In the remainder of this subsection, we pretend, for simplicity, that $T_0 = 6\varepsilon^2$ and $T_{\pm 2} = 3\varepsilon^2$.

We define the multiplication operators K_n by

$$K_n = \alpha - (1-(\kappa+\omega n)^2)^2 - T_0 - \sigma .$$

Then, for $n = \pm 1$, Eq. (19.9) takes the approximate form

$$\begin{aligned} K_1 w_1 - T_2 w_{-1} &= f_1 , \\ K_{-1} w_{-1} - T_{-2} w_1 &= f_{-1} . \end{aligned} \tag{19.13}$$

Therefore, the action of A_κ is approximated by that of the matrix

$$\mathbf{A}_\kappa = \begin{pmatrix} K_1 & -T_2 \\ -T_{-2} & K_{-1} \end{pmatrix} . \tag{19.14}$$

For $\kappa = 0$, the matrix $\mathbf{A}_\kappa$ reduces to

$$\mathbf{A}_0 = -3\varepsilon^2 \begin{pmatrix} 1 & 1 \\ 1 & 1 \end{pmatrix} .$$

Thus, we expect that the operator $A_{\kappa=0}$ has spectrum near 0 and near $-6\varepsilon^2$. In fact, we have already seen from the translation invariance of the SH equation that $A_{\kappa=0}$ has an eigenvalue 0. The corresponding eigenvector is $(1,-1)$ (odd functions), while for the eigenvalue $-6\varepsilon^2$ it is $(1,1)$ (even functions).

We now compute the shape of the spectrum of $\mathbf{A}_\kappa$ for κ close to zero. It is useful to parametrize ω and κ as follows:

$$\omega^2 = 1 + \sqrt{3}\varepsilon W , \quad \kappa = \sqrt{3}\varepsilon K/2 . \tag{19.15}$$

A straightforward calculation shows that (when $\omega > 0$)

$$(1-(\omega \pm \kappa)^2)^2 = 3\varepsilon^2 (W \pm K)^2 + \mathcal{O}(\varepsilon^3) .$$

For simplicity, we shall neglect from now on the higher order terms $\mathcal{O}(\varepsilon^3)$. Using again $\alpha = (1-\omega^2)^2 + 3\varepsilon^2$, we see that the matrix $\mathbf{A}_\kappa$ is equal to

$$3\varepsilon^2 \begin{pmatrix} W^2+1-(W+K)^2-2 & -1 \\ -1 & W^2+1-(W-K)^2-2 \end{pmatrix} .$$

The eigenvalues of the matrix are

$$\sigma_\pm = 3\varepsilon^2 \cdot \left(-K^2 - 1 \pm \sqrt{4W^2K^2+1}\right) . \tag{19.16}$$

Of course, we have $\sigma_+(K=0)=0$. Differentiating σ_+ with respect to K^2, we see that

$$\frac{d}{dK^2}\sigma_+ = 3\varepsilon^2 \cdot \left(-1 + \frac{4W^2}{2\sqrt{4W^2K^2+1}}\right) . \tag{19.17}$$

Therefore, it follows that at $\kappa = 0$,

$$\frac{d}{d\kappa^2}\sigma_+ < 0 , \quad \text{if} \quad |W| < \frac{1}{\sqrt{2}} \quad \text{i.e.,} \quad 2(1-\omega^2)^2 < 3\varepsilon^2 . \tag{19.18}$$

Similarly,

$$\frac{d}{d\kappa^2}\sigma_+ > 0 , \quad \text{if} \quad |W| > \frac{1}{\sqrt{2}} \quad \text{i.e.,} \quad 2(1-\omega^2)^2 > 3\varepsilon^2 . \tag{19.19}$$

The solutions of $\mathbf{M}X = e^\sigma X$. We reconsider the 4×4 matrix $\mathbf{M}$ defined in (19.8). It has two eigenvalues, e^σ and $e^{-\sigma}$, with eigenvectors (X_0, X_1, X_2, X_3) and $(X_0, -X_1, X_2, -X_3)$. This is seen by fixing W in (19.16) and solving for $\sigma_+ = 0$. The nontrivial solutions are

$$K = \pm\sqrt{4W^2-2}$$

and we see that they correspond to a real σ if and only if K is imaginary, that is, if $|W| < 2^{-1/2}$. Then

$$\sigma = \frac{2\pi i \kappa}{\omega} = \frac{2\pi}{\omega}\frac{\sqrt{3}\varepsilon}{2}\sqrt{2-4W^2} = \frac{\pi}{\omega}\sqrt{6\varepsilon^2 - 4(1-\omega^2)^2} . \tag{19.20}$$

Furthermore, there is always a solution with eigenvalue 0 for A (i.e., with eigenvalue 1 for $\mathbf{M}$) corresponding to the function $v(x) = u'(x)$. Since the eigenvalue 0 is double, there must be a second solution, but it is not a periodic function of x. (This is the reason why one does not find it in the Fourier analysis above.) It corresponds to the nilpotent part of $\mathbf{M}$. We construct it now explicitly. Denote $A = A_{\kappa=0}$. Since u solves (19.3), we see that the derivative u' satisfies

$$Au' = 0 .$$

We shall construct an eigenvector of the form $xu'(x)+$ periodic function. Consider first the function

$$v(x) = xu'(x) .$$

It is straightforward to check that

$$(Av)(x) = -4\partial_x^2 u(x) - 4\partial_x^4 u(x) \equiv s(x) . \tag{19.21}$$

We see that s is an *even, periodic* function of x. The symmetry $x \leftrightarrow -x$ of (19.3) implies that A_κ maps even functions to even functions and odd functions to odd functions. It is easy to see that the spectrum in the even subspace has as its highest eigenvalue the eigenvalue $-6\varepsilon^2$. That is, the eigenvalue 0 is associated with the odd subspace. Therefore, the problem

$$(Ah)(x) = -s(x) ,$$

has a unique even *periodic* solution h. If we now define $v(x) = xu'(x) + h(x)$, then we see that $Av = 0$. This solution is linearly independent of u', and thus, we have found the second solution to $A_{\kappa=0}v = 0$.

Remark. One can understand the second eigenvector with eigenvalue 0 in the following way. The set of stationary solutions of the SH equation is parametrized by ε and ω. These solutions are given by $u_{\varepsilon,\omega}(x) \approx 2\varepsilon\cos(\omega x)$. Note now that if we consider a *fixed equation*, that is,

$$\left(\alpha - (1+\partial_x^2)^2\right)u_{\varepsilon,\omega}(x) = u_{\varepsilon,\omega}^3(x) , \tag{19.22}$$

then the two parameters ε and ω are constrained by the condition

$$\alpha = \lambda(\varepsilon) + (1-\omega^2)^2 , \tag{19.23}$$

as can be seen from (17.18). Recall that $\lambda(\varepsilon) \approx 3\varepsilon^2$. Therefore, if we choose, for example, ω as the free parameter, then we see that (19.22) has the solutions

$$u_{\varepsilon,\omega}(x) \approx \frac{2\sqrt{\alpha - (1-\omega^2)^2}}{\sqrt{3}} \cos(\omega x) . \tag{19.24}$$

The problem has 2 symmetries: translation invariance and invariance with respect to ω. If we differentiate with respect to x, we obtain u' as above and if we differentiate with respect to ω, we obtain a function which is approximated by

$$\partial_\omega u_{\varepsilon,\omega}(x) \approx x\partial_x u_{\varepsilon,\omega}(x) + \cos(\omega x) \cdot \partial_\omega \frac{2\sqrt{\alpha - (1-\omega^2)^2}}{\sqrt{3}} .$$

This explains why the second eigenvector has the form we found above.

20. Linear Stability Analysis for Stationary Solutions

In this section, we study the stability of the bifurcated stationary solutions under infinitesimal perturbations. In other words, we consider now the bifurcated solutions from Section 17.2 as the basic solutions and ask about their stability with respect to infinitesimal perturbations. We consider only perturbations which are fixed in the laboratory frame. To study these, we concentrate again on the Swift-Hohenberg equation

$$\partial_t U(x,t) = \alpha U(x,t) - (1+\partial_x^2)^2 U(x,t) - U(x,t)^3 . \tag{20.1}$$

We fix $\alpha > 0$ and consider all frequencies ω satisfying $\alpha > (1-\omega^2)^2$. The arguments of this section will apply for small α only. According to the theory of Section 17.2, the equation (20.1) has – for α, ω in the above range – a stationary solution $U(x,t) = u(x)$ which is periodic with period $2\pi/\omega$. It is the *linear stability* of this solution which we want to analyze. We shall show that these solutions are *stable* for

$$(1-\omega^2)^2 < \frac{\alpha}{3}$$

and *unstable* for

$$\frac{\alpha}{3} < (1-\omega^2)^2 < \alpha .$$

This instability is called the *Eckhaus instability*. One can reformulate this dichotomy in terms of the bifurcation parameter ε:* $3\varepsilon^2 = \alpha - (1-\omega^2)^2$. Then, stability for fixed ω occurs for

$$3\varepsilon^2 > 2(1-\omega^2)^2 , \tag{20.2}$$

and instability for

$$3\varepsilon^2 < 2(1-\omega^2)^2 .$$

We recall that the factor 3 is a consequence of our particular normalization: We write the bifurcating solution as $2\varepsilon\cos(\omega x)$ +higher order terms.

Remark. Usually, the Eckhaus instability is represented in a diagram in the ω, α plane, see Fig. 13. From the point of view of bifurcation theory, ω and ε are the natural variables. The lower curve is a piece of $\alpha = (1-\omega^2)^2$. It is the threshold for the existence of solutions. Solutions exist above the curve for small ε. The upper curve is $\alpha/3 = (1-\omega^2)^2$ which is the threshold for Eckhaus instability. Solutions are stable above this curve. In Theorem 20.1, we formulate the instability in terms of the variables familiar to physicists.

As explained in Section 19.1, the linearized equation is

$$\partial_t v(x,t) = \alpha v(x,t) - (1+\partial_x^2)^2 v(x,t) - 3u(x)^2 v(x,t) . \tag{20.3}$$

Since u is a function which is periodic in x, one can view (20.3) as an evolution equation for a function v in a periodic external potential $u(x)$. We are interested in "eigenmodes" for this equation which are stable or unstable. If they are all stable, that is, if the spectrum of the evolution operator (20.3) lies in the half-plane $\mathrm{Re}\, z < 0$, then we say that u is stable under infinitesimal perturbations. If the spectrum intersects the half-plane $\mathrm{Re}\, z > 0$, then u is called unstable.

According to the theory of Bloch waves [RS], this kind of instability can be studied in terms of the eigenfunctions of the operator on the r.h.s. of (20.3) which are of the form $v(x) = w(x)e^{i\kappa x}$, where w is a $2\pi/\omega$-periodic function and $\kappa \in [-\omega/\pi, \omega/\pi]$. In other words, we study the spectrum of the operator A_κ, given by

$$A_\kappa w(x) = \alpha w(x) - (1-(\kappa - i\partial_x)^2)^2 w(x) - 3u(x)^2 w(x) .$$

and acting on $w \in \mathrm{L}^2_{\mathrm{per}}([0, 2\pi/\omega], dx)$.

* This bifurcation parameter differs from the one in Section 17.2 by higher order terms.

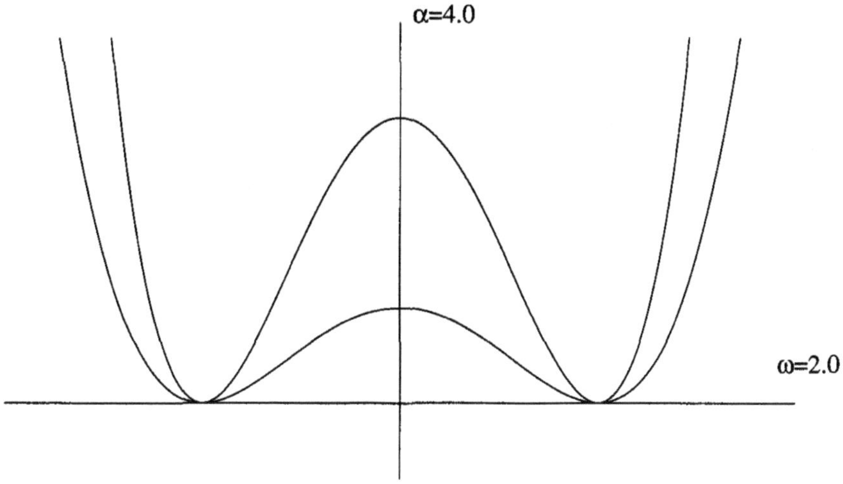

Fig. 13: The Eckhaus instability in the ω, α plane. Solutions exist above the lower curve and are stable above the upper curve.

Theorem 20.1. *For sufficiently small α the following is true: For every ω satisfying*

$$|\omega^2 - 1| < (\alpha/3)^{1/2} , \tag{20.4}$$

and for every κ satisfying $-\omega/\pi \leq \kappa \leq \omega/\pi$ the spectrum of A_κ on $L^2_{\text{per}}([0, 2\pi/\omega])$ lies in the closed left half-plane. On the other hand, if the inequality (20.4) is reversed, the spectrum of A_κ intersects the right half-plane for some κ.

Definition. The kind of instability described above is called Eckhaus instability [E].

In order to study the spectrum of A_κ, we analyze the existence of solutions of

$$(A_\kappa - \sigma)w = f \tag{20.5}$$

when $f \in L^2_{\text{per}}([0, 2\pi/\omega])$. If the Eq. (20.5) has a unique solution $w \in L^2_{\text{per}}([0, 2\pi/\omega])$, then σ is not an eigenvalue of A_κ.

Theorem 20.2. *Let $a < \sqrt{3/2}$, and assume that $|\omega^2 - 1| \leq a\varepsilon$. For* *sufficiently small ε (depending on a), the operator $A_\omega - \sigma$ has a bounded inverse on $L^2_{\mathrm{per}}([0, 2\pi/\omega])$ for σ in the region* **D** *of Fig. 14.*

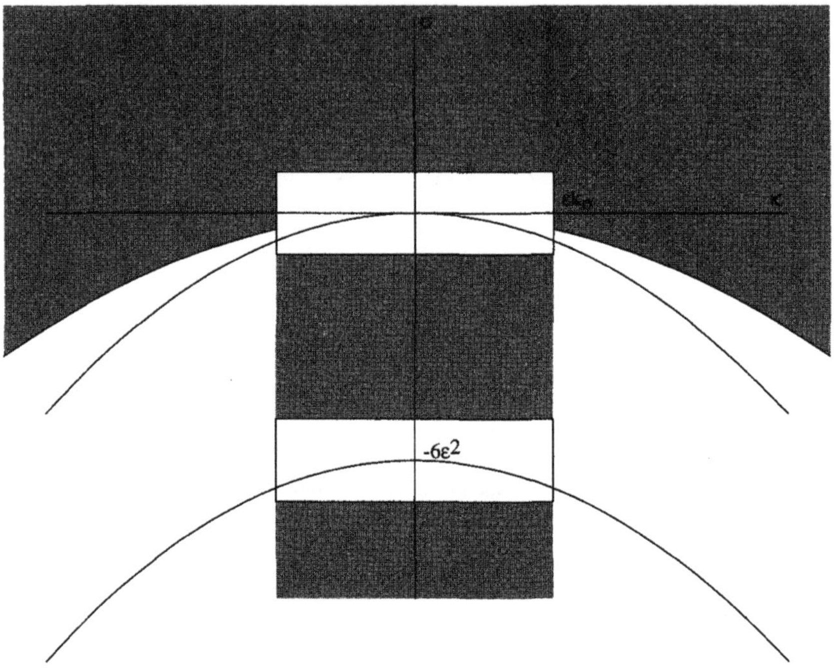

Fig. 14: For every κ, the vertical axis is the spectrum of A_κ. The shaded region, **D**, is known to contain no spectrum. The two parabolas show the typical shape of the spectrum when the system is stable.

20.1. Perturbation Theory

We have already done perturbation theory for A_κ in Section 19.2. Now we interpret the results. We found that, in perturbation theory, A_κ has two simple eigenvalues with eigenvectors in $L^2_{\mathrm{per}}([-\pi/\omega, +\pi/\omega])$, namely $\sigma_+ = 0$ and $\sigma_- \approx -6\varepsilon^2$. Also, we have seen that when $\kappa = 0$,

$$\frac{d}{d\kappa^2}\sigma_+ < 0\,, \quad \text{if} \quad |W| < \frac{1}{\sqrt{2}} \quad \text{i.e.,} \quad 2(1-\omega^2)^2 < 3\varepsilon^2\,. \tag{20.6}$$

Similarly,

$$\frac{d}{d\kappa^2}\sigma_+ > 0\,, \quad \text{if} \quad |W| > \frac{1}{\sqrt{2}} \quad \text{i.e.,} \quad 2(1-\omega^2)^2 > 3\varepsilon^2\,. \qquad (20.7)$$

Therefore, the spectrum is of one of the two types of Fig. 15.

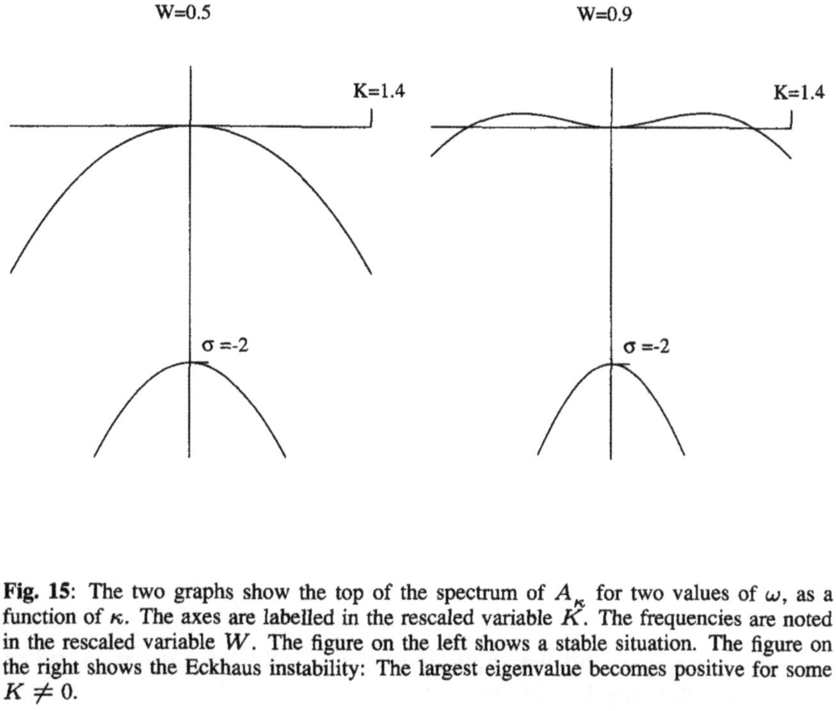

Fig. 15: The two graphs show the top of the spectrum of A_κ for two values of ω, as a function of κ. The axes are labelled in the rescaled variable K. The frequencies are noted in the rescaled variable W. The figure on the left shows a stable situation. The figure on the right shows the Eckhaus instability: The largest eigenvalue becomes positive for some $K \neq 0$.

In this section, we are interested in *real* κ, that is, in *bounded* solutions. (In Section 21 we will be interested in *imaginary* κ, that is, in *exponentially decaying*, or increasing, solutions.) We find that for (small) real κ, that is, for $\kappa^2 > 0$, the quantity σ_+ will take positive values if $|W| > 2^{-1/2}$. The operator A_κ will then have spectrum in the right half-plane, indicating an instability to linear perturbations. This is again the *Eckhaus instability*.

20.2. Proof of Theorem 20.2

Our basic strategy is to invert the operator on the l.h.s. of Eq. (19.10) and to view the problem Eq. (20.5) as a contraction problem. The proof will be in two parts, but follows the perturbative argument quite closely. We begin by showing the *absence* of spectrum in certain regions of the κ, ω-plane. In particular, we shall see that for small κ there are two isolated eigenvalues close to zero. We shall then use perturbation theory to show that these two eigenvalues are nonpositive. Several modifications of this basic strategy will be needed, and in particular, we need to consider with special care the sectors $n = +1, -1$ and $n = 0, \pm 2$. We have already defined the multiplication operators K_n by

$$K_n = \alpha - (1 - (\kappa + \omega n)^2)^2 - T_0 - \sigma .$$

We now define the (convolution) operators $B_{n,j}$, by

$$w \mapsto B_{n,j}(w) = 3 \sum_{\substack{|p+q| \geq j \\ p+q+r=n}} u_p u_q w_r .$$

With these notations, Eq. (20.5) takes the form

$$K_n w_n = B_{n,1}(w) + f_n . \tag{20.8}$$

We define the map $\mathcal{U}_n$ by

$$\mathcal{U}_n(w) = K_n^{-1} B_{n,1}(w) + K_n^{-1} f_n .$$

This map is linear, and we shall show that (20.8) has a solution, when $|n| \neq 1$, by showing that $\mathcal{U}_n$ is a contraction.

We begin by giving bounds on $B_{n,j}$. We have seen in Theorem 17.1 that u satisfies the bound

$$|u_n| < \begin{cases} \text{const. } \varepsilon & \text{for } |n| = 1, \\ \text{const. } \varepsilon^2 \rho^{-|n|} & \text{for } |n| \neq 1. \end{cases} \tag{20.9}$$

Therefore, there is a constant L such that for all $j \geq 0$, one has

$$|B_{n,j}(w)| \leq L\varepsilon^2 \|w\|_2 . \tag{20.10}$$

Here, $\|\cdot\|_2$ is the l^2-norm, that is, we use a notation different from Section 17.3. The inequality (20.10) follows from

$$\begin{aligned} |B_{n,j}(w)| &\leq \mathcal{O}(\varepsilon^2)\sum_r |w_r| \sum_{p+q=n-r} \rho^{-(|p|+|q|)} \\ &\leq \mathcal{O}(\varepsilon^2)\sum_r |w_r|\rho^{-|n-r|} \\ &\leq \mathcal{O}(\varepsilon^2)\|w\|_2 \ . \end{aligned} \tag{20.11}$$

Eq. (20.9) implies that for $j \geq 3$ one has the better bound

$$|B_{n,j}(w)| \leq L\varepsilon^3\|w\|_2 \ . \tag{20.12}$$

We begin now the detailed study of Eq. (20.8).

Case $|n| \geq 3$. In this case, we write Eq. (20.8) as

$$w_n = K_n^{-1}B_{n,1}(w) + K_n^{-1}f_n \ . \tag{20.13}$$

Since $|\kappa| \leq \omega/\pi$ and $T_0 > 0$, we see that the operator $-K_n$ is bounded below by

$$-K_n \geq Cn^4,$$

for all $\sigma \geq -1$, for some $C > 0$. Hence,

$$|K_n^{-1}f_n| \leq C^{-1}n^{-4}\|f\|_2 \ ,$$

and

$$|K_n^{-1}B_{n,1}(w)| \leq \mathcal{O}(\varepsilon^2 n^{-4})\|w\|_2 \ ,$$

so that

$$\left(\sum_{|n|\geq 3} |K_n^{-1}B_{n,1}(w)|^2\right)^{1/2} \leq \mathcal{O}(\varepsilon^2)\|w\|_2 \ .$$

Thus, the homogeneous part of $\mathcal{U}_n$ is a contraction when $|n| \geq 3$.

Case $|n| = 2$ or $n = 0$. For $|n| = 2$, resp. $n = 0$, the operators K_n take the form

$$\begin{aligned} K_{\pm 2} &= \alpha - (1 - (\kappa \pm 2\omega)^2)^2 - T_0 - \sigma \ , \\ K_0 &= \alpha - (1 - \kappa^2)^2 - T_0 - \sigma \ . \end{aligned}$$

Since $|\kappa| \le \omega/\pi$, we see that for $n = -2, 0, 2$, one has

$$|K_n^{-1} f_n| \le \mathcal{O}(1)\|f\|_2 .$$

Case $|n| = 1$. This is really the interesting case, because the linear operator has spectrum very close to zero. In fact, for $\kappa = 0$, zero is in the spectrum of A_κ. We study this case as a coupled system of equations, which we first rewrite in the form

$$\begin{aligned} K_1 w_1 - T_2 w_{-1} &= T_{-2} w_3 + B_{1,4}(w) + f_1 , \\ -T_{-2} w_1 + K_{-1} w_{-1} &= T_2 w_{-3} + B_{-1,4}(w) + f_{-1} . \end{aligned}$$

In matrix notation, this is then written as

$$\mathbf{A}\begin{pmatrix} w_1 \\ w_{-1} \end{pmatrix} = \begin{pmatrix} T_{-2} w_3 \\ T_2 w_{-3} \end{pmatrix} + \begin{pmatrix} B_{1,4}(w) \\ B_{-1,4}(w) \end{pmatrix} + \begin{pmatrix} f_1 \\ f_{-1} \end{pmatrix} , \tag{20.14}$$

with

$$\mathbf{A} = \begin{pmatrix} K_1 & -T_2 \\ -T_{-2} & K_{-1} \end{pmatrix} \tag{20.15}$$

$$= \begin{pmatrix} \alpha - (1 - (\omega + \kappa)^2)^2 - T_0 - \sigma & -T_2 \\ -T_{-2} & \alpha - (1 - (\omega - \kappa)^2)^2 - T_0 - \sigma \end{pmatrix} .$$

Instead of solving Eq. (20.14), we use the identity Eq. (20.13), and we reformulate the problem as

$$\begin{aligned} \begin{pmatrix} w_1 \\ w_{-1} \end{pmatrix} &= \mathbf{A}^{-1} \begin{pmatrix} T_{-2} K_3^{-1} B_{3,1}(w) \\ T_2 K_{-3}^{-1} B_{-3,1}(w) \end{pmatrix} + \mathbf{A}^{-1} \begin{pmatrix} B_{1,4}(w) \\ B_{-1,4}(w) \end{pmatrix} \\ &+ \mathbf{A}^{-1} \begin{pmatrix} f_1 + K_3^{-1} f_3 \\ f_1 + K_{-3}^{-1} f_{-3} \end{pmatrix} \\ &\equiv \mathbf{A}_1(w) + \mathbf{A}_1' \mathbf{f} . \end{aligned} \tag{20.16}$$

In this way, we shall gain a factor ε^2 in the operator $\mathbf{A}_1$. We call this the technique of improving inhomogeneities. We view the r.h.s. of Eq. (20.16) as a map from l^2 to itself. To control this map, we first study the inverse of $\mathbf{A}$. For sufficiently small κ, the eigenvalues of $\mathbf{A}$ are

$$-6\varepsilon^2 \pm 3\varepsilon^2 + \alpha - (1 - \omega^2)^2 - \sigma + \mathcal{O}(\kappa^2) + o(\varepsilon^2) \equiv X .$$

Note that, by definition, $3\varepsilon^2 = \alpha - (1-\omega^2)^2$, so that

$$X = -3\varepsilon^2 \pm 3\varepsilon^2 - \sigma + \mathcal{O}(\kappa^2) + o(\varepsilon^2) .$$

If we consider the "+" sign, then we see that $|X| > \varepsilon^2/4$, provided

$$|\sigma| > \varepsilon^2/2 \quad \text{and} \quad |\kappa| \le \varepsilon k_0 , \tag{20.17}$$

with k_0 sufficiently small. If we consider the "−" sign, then we see that $|X| > \varepsilon^2/4$, provided

$$|\sigma + 6\varepsilon^2| > \varepsilon^2/2 \quad \text{and} \quad |\kappa| \le \varepsilon k_0 , \tag{20.18}$$

with k_0 sufficiently small. Thus, if (20.17) and (20.18) hold, then $\|\mathbf{A}^{-1}\| \le 4/\varepsilon^2$, because $\mathbf{A}$ is symmetric. This means that Eq. (20.16) is defined in this case. Note now that the first two terms on the r.h.s. of Eq. (20.16) are maps whose norms are bounded (using (20.10) and (20.12)) by

$$(4/\varepsilon^2)\cdot \varepsilon^2 \cdot \mathcal{O}(1)\varepsilon^2 + (4/\varepsilon^2)\cdot \mathcal{O}(1)\varepsilon^3 , \tag{20.19}$$

and this is smaller than 1 if ε is sufficiently small. Hence they are contractions and Eq. (20.16) has a unique solution in this case. We have therefore shown: There is no spectrum in the regions described by (20.17) and (20.18).

We next consider the region $|\kappa| > \varepsilon k_0$. As in the perturbative argument, it is useful to parametrize ω and κ as follows:

$$\omega^2 = 1 + \sqrt{3}\varepsilon W , \quad \kappa = \sqrt{3}\varepsilon K/2 . \tag{20.20}$$

Then we obtain, as before, but now with the definition (19.11),

$$(1-(\omega \pm \kappa)^2)^2 = 3\varepsilon^2 (W \pm K)^2 + \mathcal{O}(\varepsilon^3) .$$

We can rewrite the matrix $\mathbf{A} + \sigma\mathbf{1}$ as

$$3\varepsilon^2 \begin{pmatrix} W^2+1-(W+K)^2-2 & -1 \\ -1 & W^2+1-(W-K)^2-2 \end{pmatrix} + \mathcal{O}(\varepsilon^3).$$

The determinant of $(3\varepsilon^2)^{-1}(\mathbf{A}+\sigma\mathbf{1})$ is

$$(K^2+2KW+1)(K^2-2KW+1)-1 = K^2(K^2-4W^2+2) , \tag{20.21}$$

and this quantity is nonnegative for all κ if and only if

$$|W| \leq \frac{1}{\sqrt{2}} . \tag{20.22}$$

If the inequality in (20.22) is strict, then both eigenvalues of

$$(3\varepsilon^2)^{-1}(\mathbf{A} + \sigma \mathbf{1})$$

have the same sign (in fact, they are negative). The two eigenvalues are

$$\sigma_{\pm} = -K^2 - 1 \pm \sqrt{4W^2K^2 + 1} ,$$

and so one eigenvalue satisfies $\sigma_- < -2$ and the other

$$\sigma_+ < -\text{const. } K^2(1 - 2W^2) .$$

In the Theorem 20.2, we *assume* that $|W| \leq a < 2^{-1/2}$, and therefore $(3\varepsilon^2)^{-1}(\mathbf{A} + \sigma\mathbf{1})$ has no spectrum for $\sigma \geq -\mathcal{O}(1)K^2\varepsilon^2$, which means that $\mathbf{A}$ is invertible for $\sigma \geq -\mathcal{O}(1)K^2\varepsilon^2$. For σ in this region the inverse of $\mathbf{A}$ exists when $|W| \leq a < 2^{-1/2}$, and is bounded by

$$\mathcal{O}(\varepsilon^{-2}k_0^{-2}(1 - 2W^2)^{-1}) .$$

It follows, as before in Eq. (20.19), that for sufficiently small ε – the smallness depends on a – the operator occurring in Eq. (20.16) is a contraction, if $\sigma \geq -\text{const. } K^2\varepsilon^2$. This shows that there is no spectrum in $\sigma \geq -\text{const. } K^2\varepsilon^2$, when $|\kappa| > \varepsilon k_0$.

If (20.22) is violated, then one can find an unstable state (for sufficiently small α). We leave it to the reader to fill in the details of this argument. This completes the proof of Theorem 20.2.

At this point, we have shown that there is no spectrum in the hatched region of Fig. 14. It remains to show that in the central part, that is, in the region $|\kappa| < k_0\varepsilon$, there is no spectrum above zero. This will then show the stability of the stationary solution and complete the proof of Theorem 20.1. The method of proof is easy, because by the results of Theorem 20.2, the largest eigenvalue is simple, and we can apply ordinary perturbation theory (cf. [K]). We have already done a perturbation analysis in Section 19.2, but now we use a somewhat more general formalism, which should be useful for other problems as well.

By definition, we can expand A_κ in powers of κ, and we have

$$A_\kappa = \sum_{m=0}^{4} A^{(m)}\kappa^m .$$

Since $A^{(k)}$ is relatively bounded in the space of periodic functions with respect to $A^{(0)}$, for $k = 1, \dots 4$, we can apply analytic perturbation theory in κ. We get for the smallest eigenvalue σ_κ an expansion

$$\sigma_\kappa = \sum_{m=0}^{\infty} \sigma^{(m)}\kappa^m .$$

If we denote by $e^{(j)}$ the coefficients of the expansion of the eigenvector, then the first few coefficients satisfy the equations

$$\begin{aligned} A^{(0)}e^{(0)} &= \sigma^{(0)}e^{(0)} , \\ A^{(0)}e^{(1)} + A^{(1)}e^{(0)} &= \sigma^{(0)}e^{(1)} + \sigma^{(1)}e^{(0)} , \\ A^{(0)}e^{(2)} + A^{(1)}e^{(1)} + A^{(2)}e^{(0)} &= \sigma^{(0)}e^{(2)} + \sigma^{(1)}e^{(1)} + \sigma^{(2)}e^{(0)} . \end{aligned}$$

These formulas are valid, because we have seen that the kernel of $A^{(0)}$ has dimension 1. Note now that $\sigma^{(0)} = 0$ and $e^{(0)} = \partial_x u(x)$ because the space coordinate does not appear explicitly in (20.1). We have

$$\begin{aligned} \sigma^{(1)}(e^{(0)}, e^{(0)}) &= (e^{(0)}, A^{(0)}e^{(1)}) + (e^{(0)}, A^{(1)}e^{(0)}) \\ &= 0 + 0 . \end{aligned}$$

The notation $(\cdot, \cdot)$ stands for the scalar product in L^2. The first zero above is a consequence of the symmetry of $A^{(0)}$, and the second follows by observing that $A^{(1)}$ has an odd number of derivatives and $f_n = f_{-n}$, since f is real. Therefore, $\sigma^{(1)} = 0$, and we get

$$\sigma^{(2)} = \frac{1}{(e^{(0)}, e^{(0)})} \left((e^{(0)}, A^{(1)}e^{(1)}) + (e^{(0)}, A^{(2)}e^{(0)})\right) . \qquad (20.23)$$

The term $(e^{(0)}, A^{(0)}e^{(1)})$ is absent because $A^{(0)}$ is symmetric. From

$$A^{(0)}e^{(1)} = -A^{(1)}e^{(0)}$$

we deduce

$$e^{(1)} = -\left(A^{(0)}\right)^{-1} A^{(1)}e^{(0)} + \rho e^{(0)} , \qquad (20.24)$$

where the value of ρ is irrelevant for what follows. We have to show that $A^{(0)}$ is invertible, but since it has a potentially degenerate eigenvalue, we only show that the equation

$$A^{(0)}w = f \tag{20.25}$$

has a unique solution provided we require $w_1 = w_{-1}$. Then the problem (20.25) is really a special case of our previous analysis: For $|n| \neq 1$, we view (20.25) as the fixed point problem (20.13),

$$w_n = \frac{B_{n,1}(w) + f_n}{3\varepsilon^2 - (1 - n^2\omega^2)^2 - T_0}, \tag{20.26}$$

and the operator on the r.h.s. of (20.26) is a contraction. Now if $w_1 = w_{-1}$, then the matrix **A** of (20.15) becomes multiplication by

$$X = \alpha - (1 - \omega^2)^2 - T_0 - T_2 \tag{20.27}$$

so that we can write the $n = 1$ component of (20.25) as

$$\begin{aligned} w_1 &= X^{-1}\left(T_{-2}w_3 + B_{1,4}(w) + f_1\right) \\ &= X^{-1}\left(T_{-2}K_3^{-1}B_{3,1}(w) + B_{1,4}(w)\right) + X^{-1}(f_1 + K_3^{-1}f_3) . \end{aligned}$$

Again, the operator on the r.h.s. is a contraction and thus the proof of invertibility of $A^{(0)}$ is complete.

We now compute the second derivative $\sigma^{(2)}$ of σ_κ at $\kappa = 0$, showing that it has a negative sign. We have

$$(e^{(0)}, A^{(1)}e^{(1)}) = -(e^{(0)}, A^{(1)}\left(A^{(0)}\right)^{-1}A^{(1)}e^{(0)}) . \tag{20.28}$$

So all the terms exist, and perturbation theory is valid. We have

$$\begin{aligned} A^{(0)} &= \alpha - (1 + \partial_x^2)^2 - 3u(x)^2 , \\ A^{(1)} &= -4i(\partial_x + \partial_x^3) , \\ A^{(2)} &= 2 + 6\partial_x^2 . \end{aligned}$$

Since $e^{(0)}$ is proportional to $\sin(\omega x)$, we find from (20.23), (20.28), (20.27) and (19.12), neglecting higher order terms in ε,

$$\begin{aligned} \sigma^{(2)} &= -\frac{\left(4\omega(1 - \omega^2)\right)^2}{X} + 2 - 6\omega^2 \\ &= -\frac{16\omega^2 \cdot 3\varepsilon^2 W^2}{3\varepsilon^2 - 6\varepsilon^2 - 3\varepsilon^2} + 6\sqrt{3}\varepsilon W - 4 \\ &= -4 + 8W^2 + \mathcal{O}(\varepsilon) . \end{aligned} \tag{20.29}$$

This quantity is negative if $|W| < 2^{-1/2}$, and ε is sufficiently small. We see again the appearance of the Eckhaus instability: It indicates the parameter values where the highest eigenvalue crosses zero.

We can now complete the proof of Theorem 20.1. Near $\kappa = 0$, we see from Theorem 20.2 that perturbation theory holds. Then (20.29) implies that for sufficiently small $|\kappa|$, the spectrum lies below zero. This dictates a (possibly smaller) choice of k_0 in the proof of Theorem 20.2. Now Theorem 20.2 applies for sufficiently small ε (depending on this k_0 and the value of ω). Since we have already shown that there is no spectrum in the region **D** of Fig. 14, the assertion of the theorem follows.

20.3. Stability Analysis of Bifurcated Periodic Solutions

We have analyzed the stability of stationary solutions for the particular example of the SH equation. We now reformulate these results in terms of a more general method to analyze the stability of bifurcated periodic solutions. Examples of such bifurcations were discussed in detail for the SH equation and the convective bifurcations.

It is technically simpler to analyze the spectrum of the linearized operator in the space $\mathrm{L}^2(\mathbf{R}, dx)$ of square integrable functions, and we shall first deal with this question. We shall then show that the spectrum in the space $\mathrm{L}^\infty(\mathbf{R}, dx)$ of bounded functions is the same (we recall that according to Section 4, this is the natural space to work with).

We shall show that under suitable hypotheses (natural for bifurcation problems), the question of stability can be reduced to a problem of perturbation theory for a finite number of eigenvalues. In a concrete case, one has to verify the hypotheses below and then to compute the perturbation theory to some finite order.

Note also that proving the stability of the bifurcated solution in a space of functions with the same period is a much easier task but is irrelevant to the present discussion.

As explained in Section 19, since the linearized operator around the bifurcated solution has periodic coefficients (with period $2\pi/\omega$), one can use a Bloch wave technique to reduce the problem to a family of spectral problems on the interval $[0, 2\pi/\omega]$. More precisely, if A_ε denotes the tangent map around the bifurcated solution (of period $2\pi/\omega$), the Bloch operators $A_{\varepsilon,\kappa}$ for $\kappa \in [0, \omega]$ are given by

$$A_{\varepsilon,\kappa} f(x) = e^{-i\kappa x} A_\varepsilon e^{i\kappa x} f(x) .$$

One can give an expression (formal at this point) for the solution of the equation

$$(A_\varepsilon - z)h = f$$

when z is not in the spectrum of A_ε :

$$h(x) = \int_0^\omega d\kappa \int_0^{2\pi/\omega} dy\, e^{-\imath\kappa x}$$
$$(A_{\varepsilon,\kappa} - z)^{-1}(x,y)\, e^{i\kappa y} \sum_{n\in\mathbf{Z}} e^{2\pi i\kappa n/\omega} f(y + 2\pi n/\omega)\ .$$

We have used the notation $K(x,y)$ for the integral kernel of the operator K.

We shall now make several hypotheses inspired by the bifurcation problem. We shall assume that these hypotheses are valid in L^2 spaces.

H1. *The difference $A_{\varepsilon,\kappa} - A_{0,\kappa}$ is relatively bounded with respect to $A_{0,\kappa}$ uniformly in κ, and the relative bound is of the order of a power of ε. In other words, there are two positive constants C, and $p \geq 1$, such that for ε small enough and uniformly in κ, for any vector u in the domain of $A_{0,\kappa}$ we have*

$$\|A_{\varepsilon,\kappa}u - A_{0,\kappa}u\| \leq C\varepsilon^p(\|A_{0,\kappa}u\| + \|u\|)\ .$$

We note that the generic value of p is 1, although it frequently takes the value 2 for the global stability of the equation (see for example the case of the SH equation, where one can add a quadratic term). One expects p to depend on the nature of the bifurcation.

H2. *The spectrum of $A_{0,\kappa}$ is contained in the closed left half-plane, and touches the imaginary axis only for $\kappa = 0$ and $\kappa = \omega$. Moreover, there are two positive constants D and q such that*

$$\sup_{\theta\in\mathbf{R}} \|(A_{0,\kappa} - i\theta)^{-1}\| \leq D(\kappa^{-q} + (\omega - \kappa)^{-q})\ .$$

We observe that in the case where the operator A_0 has constant coefficients, the generic value of q is 2.

H3. *The operator $A_{\varepsilon,\kappa}$ is a polynomial in κ, and the coefficient of each monomial is relatively bounded with respect to $A_{\varepsilon,0}$ uniformly in ε for small ε. We shall also assume that the operator $A_{\varepsilon,\kappa}$ is* *infinitely differentiable in ε, with each derivative (and each Taylor remainder) relatively bounded with respect to $A_{0,0}$.*

This last hypothesis can, of course, be relaxed in many interesting ways. What we shall really use below is the existence of an expansion in ε and κ near $(0,0)$ with coefficients relatively bounded with respect to $A_{0,0}$.

It is easy to verify that the examples of the SH equation and the convective bifurcations satisfy the above three hypotheses, with $p = 2$.

The following result is a straightforward consequence of the above hypotheses.

Theorem 20.3. *Under the hypotheses H1 – H3, there is a positive number β such that for ε small enough,*

$$\sup_{\kappa \in (\beta\varepsilon^{p/q}, \omega - \beta\varepsilon^{p/q})} \sup_{\theta \in \mathbf{R}} \|(A_{\varepsilon,\kappa} - i\theta)^{-1}\|$$

is finite.

This implies that we only have to consider values of κ of the form $\kappa = \varepsilon^{p/q}K$ or $\kappa = \omega - \varepsilon^{p/q}K$ with $0 \leq K \leq \beta$.

We now observe that since the operator A is elliptic, the operator $A_{0,0}$ has compact resolvent. It follows from the above result and Hypothesis H3 that for ε small enough and κ as above, the only eigenvalues of $A_{\varepsilon,\kappa}$ which can possibly lie in the right half-plane are coming from perturbation of the eigenvalue(s) of $A_{0,0}$ which are on the imaginary axis. In particular, if there is no such eigenvalue, the above Bloch formula for the resolvent implies that A_{ε}, considered as an operator in $\mathrm{L}^2(\mathbf{R}, dx)$, has no spectrum in the right half-plane.

As explained at the beginning of this section, the above analysis works in L^2 spaces but we need the result in L^∞ spaces.

Theorem 20.4. *Assume that the elliptic periodic linear operator A (of period $2\pi/\omega$) is such that the Bloch operators A_κ are twice differentiable in κ with derivatives relatively bounded with respect to A_κ. Then the spectrum in $\mathrm{L}^2(\mathbf{R}, dx)$ is equal to the spectrum in $\mathrm{L}^\infty(\mathbf{R}, dx)$.*

Proof. We can write the Bloch formula for the resolvent as follows. If the complex number z is not in the spectrum, the relation $h = (A - z)^{-1} f$

can be rewritten as follows: Fix $x \in \mathbf{R}$ and let $n_x = [\omega x/2\pi]$, where $[\cdot]$ denotes the integer part. Then

$$\begin{aligned} h(x) = & \int_0^\omega d\kappa \int_0^{2\pi/\omega} dy\, e^{-i\kappa x} \\ & (A_\kappa - z)^{-1}(x,y)\, e^{i\kappa y} \sum_{|n-n_x|\le 1} e^{2\pi i\kappa n/\omega} f(y+2\pi n/\omega) \\ & + \int_0^\omega d\kappa \int_0^{2\pi/\omega} dy\, e^{-i\kappa x} \\ & (A_\kappa - z)^{-1}(x,y)\, e^{i\kappa y} \sum_{|n-n_x|>1} e^{2\pi i\kappa n/\omega} f(y+2\pi n/\omega) \;. \end{aligned}$$

The terms in the first integral pose no problem, and in the terms on the second line, we can integrate twice by parts with respect to κ, using the identity

$$e^{-i\kappa(x-2\pi n/\omega)} = \frac{\partial_\kappa e^{-i\kappa(x-2\pi n/\omega)}}{-i\kappa(x-2\pi n/\omega)} \;.$$

It is now easy to verify that due to the periodicity the boundary terms are zero, and moreover, all the terms are uniformly bounded by a constant times the convergence factor $x - 2\pi n/\omega$. The Bloch formula now makes sense provided we can show that the operators $(A_\kappa - z)^{-1}$ are bounded on the right spaces. This is, however, easy because $\mathrm{L}^\infty([0,2\pi/\omega],dx)$ is a closed subset of $\mathrm{L}^2([0,2\pi/\omega],dx)$, and the operators $(A_\kappa - z)^{-1}$ are bounded operators from $\mathrm{L}^2([0,2\pi/\omega],dx)$ to $\mathrm{L}^\infty([0,2\pi/\omega],dx)$, uniformly in κ.

We shall now briefly work out the case of convective bifurcations discussed in Section 17.4. Consider the operator of (17.30). The tangent map to this operator is of the form

$$\begin{pmatrix} \partial_x^6 - 2\partial_x^3 + c\partial_x + \alpha - 1 - 3u_1^2(x) & -5\partial_x^3 \\ \partial_x^3 & \partial_x^6 + c\partial_x + \alpha - 1 - 3u_2^2(x) \end{pmatrix} .$$

This operator is similar to the one in (17.31), except for the "potential" term coming from u_1^2 and u_2^2. Therefore, the calculations one has to do are very similar to the projection and decomposition leading to (17.37). One can first verify that for fixed ω, $\omega \neq \pm 1$, the bifurcated solutions are unstable. In this case, the constants p and q are equal to 2. The natural scaling for $\omega = \pm 1$ is ε, and one can repeat the calculation for $\omega = 1 + \varepsilon w$ (or equivalently $\omega = -1 + \varepsilon w$), where w is a fixed constant.

In this case, one finds $p = 2$, $q = 1$, and one has to perform a perturbative calculation to second order in ε. Since $q = 1$ we do the calculation for $\kappa = \varepsilon k$, with k of order 1. Since we change ω, it is advisable to rescale the space variable, so that all functions are periodic with the same period as ε is changed. The calculations are somewhat lengthy, but extremely easy to do with any computer algebra program. There are two potentially dangerous eigenvalues. One is seen to stay inside the negative half-space when $\varepsilon \neq 0$, and the other is given to order ε^2 by

$$i\varepsilon(2k - 20\varepsilon k^3(w + 2w^2)/3)$$
$$-4\varepsilon^2 k^2(2 - 5w^2) .$$

Here $k = \kappa/\varepsilon$. Thus, there arises a phenomenon which is similar to the Eckhaus instability. For $w^2 < 2/5$, the second eigenvalue is in the left half-space and therefore we see that the bifurcated solution is marginal on the stable side.

21. Bifurcation in Half-Spaces

In this section, we reconsider the bifurcation problem in the more general setting of *half-spaces*, such as $[0, \infty]$. This setting is more general because the set of solutions becomes larger: In the space $[-\infty, \infty]$, the function e^{-x} is not bounded, but on the positive half-space, it is.

Our interest in half-spaces is motivated by two main issues:

1) It is generally necessary to have a handle on the "matching problem." We want to be able to patch together solutions in half-spaces much as one can patch together solutions of *linear* problems. In other words, we want to count correctly the number of solutions on each half-space, so that we know the number of free parameters available for pasting the solutions together. For example, if the equation has a differential operator of nth order, we need n parameters to match successfully.
2) As an important application for this matching principle, we want to study bifurcation problems in *nonhomogeneous* backgrounds, for example, problems of the form

$$\big(\mu + \lambda - (1 + \partial_x^2)^2\big)u(x) = G(x, u(x)) . \qquad (21.1)$$

The cases treated in Section 17.2 correspond to a function G which depends only on the second argument: $G(x, u) = u^3$. Now we are interested in the theory of bifurcations in a situation where G may have local inhomogeneities. For example, if $G(x,u) = u^3$ for $|x| > 1$ and $G(x,u)$ is a smooth non-constant function of x and u for $|x| < 1$, with $G(x,0) = 0$. In this case, the Equation (21.1) describes a local disturbance of the nonlinear term. We would like to know how solutions bifurcate from the zero solution in this case. To study this problem, we look at the bifurcation problem in the half-spaces $[-\infty,-1]$ and $[1,\infty]$ and glue the solutions together across the disturbance. Another situation of interest occurs when $G(x,u) \to u^3$ for $x \to \infty$ and $G(x,u) \to 2u^3$ for $x \to -\infty$. This corresponds to a physical problem in which the nonlinearity has different strength at $\pm\infty$. Finally, we are also interested in the problem of different frequencies ω at $\pm\infty$. We shall not present examples for these cases.

In all of the above cases, we want to solve the problem on half-spaces and discuss then the global solution by analyzing whether or not the matching problem can be solved.

21.1. Formulation of the Problem

We look at the Swift-Hohenberg equation, written in the form

$$\alpha u(x) - (1+\partial_x^2)^2 u(x) - u^3(x) = 0 . \tag{21.2}$$

We fix ω and set $\alpha = (1-\omega^2)^2 + \lambda$ and formulate the bifurcation problem

$$\big(\lambda + (1-\omega^2)^2 - (1+\partial_x^2)^2\big)u(x) - u^3(x) = 0 ,$$

for which there is a solution of the form

$$\begin{aligned} u_0(x) &= \varepsilon\big(e^{i\omega x} + e^{-i\omega x}\big) + \mathcal{O}(\varepsilon^2) , \\ \lambda &= 3\varepsilon^2 + \mathcal{O}(\varepsilon^3) . \end{aligned}$$

In accordance with Section 17.2, we write this solution as

$$u_0(x) = \varepsilon \sum_{n\in\mathbf{Z}} \eta_n e^{in\omega x} ,$$

with $\eta_{-n} = \overline{\eta_n}$. To unify the notation, we have set $\eta_1 = \eta_{-1} = 1$. We now look for solutions on $[0,\infty]$ which are of the form

$$u(x) = u_0(x) + v(x) \tag{21.3}$$

where the amplitude v satisfies $\lim_{x\to\infty} v(x) = 0$ and $v(x) = \bar{v}(x)$. That is, we are looking for real solutions which tend to the periodic solution at infinity. Upon substituting into (21.2), we get the equation for v:

$$\left(\alpha - (1+\partial_x^2)^2 - 3u_0^2(x)\right) v(x) \; = \; 3u_0(x)v^2(x) + v^3(x) \; . \qquad (21.4)$$

We have already encountered the linear part of this equation in the study of the *stability* of u_0, cf. equation (20.3), and we have discussed its properties in Section 20. We are now interested in solutions of (21.4) on the half-space $[0, \infty]$ which go to zero as $x \to \infty$. The analysis will be based on the following strategy:

- Find the solutions to the *linear* problem corresponding to (21.4), that is, set the r.h.s. of (21.4) equal to zero and discuss the corresponding problem.
- Show that to each linear solution one can associate a nonlinear solution of the full problem. This will be done by using the theory of stable manifolds developed in Section 9. Intuitively, if the linear solution goes to zero for large x, then the nonlinear part of the problem becomes irrelevant for large x, and cannot change the nature of the problem.

21.2. The Linear Problem

We discuss the linear problem

$$\left(\alpha - (1+\partial_x^2)^2 - 3u_0^2(x)\right) v(x) \; = \; 0 \; , \qquad (21.5)$$

by using again the method of Bloch functions as in the study of stability. The operator in (21.5) is a differential operator with constant coefficients, but with a periodic "potential," namely $3u_0^2(x)$. We look for an eigenvector with eigenvalue zero for this operator. It will be of the form $v(x) = e^{i\kappa x}w(x)$, where w is a $2\pi/\omega$-periodic function of x, and $\mathrm{Im}\,\kappa > 0$, $|\mathrm{Re}\,\kappa| < \pi/\omega$. Note that in the stability analysis of Section 19.1 and Section 19.2, we had the same ansatz, but we required $\mathrm{Im}\,\kappa = 0$ since we were interested in solutions which are bounded on the whole real line. It is clear that the results of Section 19.1 carry over inasmuch as they are algebraic manipulations. To lowest order in ε in the perturbing "potential" $3u_0^2$ we obtained the following results: We parametrized ω and κ as follows:

$$\omega^2 \; = \; 1 \; + \; \sqrt{3}\varepsilon W \; , \; \kappa \; = \; \sqrt{3}\varepsilon K/2 \; . \qquad (21.6)$$

Then two eigenvalues of the operator (for fixed ω and projected onto the lowest Fourier mode) are

$$\sigma_{\pm} = -K^2 - 1 \pm \sqrt{4W^2K^2 + 1}\ .$$

We now look for the zeros of this expression, with K as the unknown, and we find

$$K^4 + 2K^2 + 1 = 1 + 4K^2W^2\ ,$$

and finally

$$K = 0\ , \quad \text{and} \quad K = \pm\sqrt{4W^2 - 2}\ . \tag{21.7}$$

We see that there are solutions with $\mathrm{Im}\,\kappa > 0$ *if and only if* $|W| < 2^{-1/2}$. This shows that exponential solutions exist exactly when the system is not Eckhaus unstable. Thus, by perturbation theory, as in Section 19.1, the linear problem (21.5) has – up to a constant factor – *exactly one* solution of the form

$$v(x) = e^{i\kappa x}w(x) \quad \text{with} \quad \mathrm{Im}\,\kappa > 0 \tag{21.8}$$

if ε is sufficiently small and

$$3\varepsilon^2 > 2(1-\omega^2)^2\ . \tag{21.9}$$

21.3. The Nonlinear Problem

We discuss this problem by viewing the ordinary differential equation (21.2) as a coupled system of 4 first order (nonlinear) differential equations. Let us denote $X(x)$ the vector whose components are the derivatives

$$X(x) = \big(v(x), \partial_x v(x), \partial_x^2 v(x), \partial_x^3 v(x)\big)\ .$$

Then we get an equation of the form

$$\partial_x X(x) = \mathbf{M}(x)X(x) + F(x, Z(x))\ , \tag{21.10}$$

where Z is the first component of the vector X,

$$\mathbf{M}(x) = \begin{pmatrix} 0 & 1 & 0 & 0 \\ 0 & 0 & 1 & 0 \\ 0 & 0 & 0 & 1 \\ 0 & -2 & 0 & \alpha - 1 - 3u_0^2(x) \end{pmatrix}$$

and $F(x, s)$ is the column vector

$$F(x,s) = (0,\ 0,\ 0,\ 3u_0(x)s^2 + s^3)\ .$$

The equation

$$\partial_x X(x) = \mathbf{M}(x)X(x) \tag{21.11}$$

has periodic coefficients. Therefore, we can define a map $\mathbf{M}$ from $\mathbf{R}^4$ to itself, by setting

$$\mathbf{M}X = X(2\pi/\omega)\ ,$$

where $X(2\pi/\omega)$ is the solution of (21.11) obtained from the initial data $X(0) = X$. In the same way, the full equation (21.10) induces a (nonlinear) "time" $2\pi/\omega$ map $\mathbf{N}$ from an initial condition in $\mathbf{R}^4$ to the resulting value one period later.

The study of exponentially decaying solutions is, in some sense, a study of *small* solutions. One expects then the nonlinear effects to become negligible. We have seen in Chapter II a full development of these ideas. Here, we only need the Theorem 9.2. This theorem says that near a zero solution, there is a manifold W^s of solutions which decay to zero. This manifold is tangent to the contracting eigenspace of the linearized problem at zero.

The situation corresponds to that described in Theorem 9.2, but with a discrete time: The problem is of the form

$$X_{n+1} = \mathbf{N}(X_n)\ ,$$

with $\mathbf{N}(X) = \mathbf{M}X + \mathbf{F}(X)$. We equip $\mathbf{R}^4$ (or $\mathbf{C}^4$) with the norm

$$\|X\| \equiv \left(\sum_{i=1}^{4} |X_i|^2\right)^{1/2}\ .$$

Then, it is easy to see from the definition of the differential equation for $X(x)$ that

$$\|\mathbf{F}(X)\| = \|\mathbf{N}(X) - \mathbf{M}X\| \le \text{const. } \|X\|^2\ .$$

This means that $\mathbf{F}$ is a nonlinear map of $\mathbf{R}^4$ to itself which satisfies the assumptions 3), (9.2) of Section 9. By our previous analysis of the linear problem, we also see that $\mathbf{M}$ has properties required by that theorem. Namely, the eigenvalues e^σ of $\mathbf{M}$ are related to the spectrum of the linear operator in (21.5) by

$$e^\sigma = e^{2\pi i\kappa/\omega}\ .$$

where κ is one of the solutions of (21.7). We can therefore apply the theory of Section 9 in a finite dimensional setting, with discrete time. Since $\mathbf{M}$ has one stable eigenvalue, described by (21.8), we have the following result:

Theorem 21.1. *For sufficiently small $\varepsilon > 0$ and for ω satisfying*

$$3\varepsilon^2 > 2(1-\omega^2)^2 ,$$

there is a one dimensional manifold $W_s \in \mathbf{R}^4$ such that $\mathbf{N}(W_s) \subset W_s$. For every $X \in W_s$, one has $\mathbf{N}^n(X) \to 0$ as $n \to \infty$. The manifold W_s is smooth.

The proof of this theorem follows at once from the proof of Theorem 9.2. The smoothness of W_s follows from the smoothness of the equations.

Remark. The Theorem 9.2 does *not* imply that the there is a 3-dimensional unstable manifold, because we are in a marginal situation, where $\mathbf{M}$ has two eigenvalues equal to one. Therefore, Theorem 9.2 implies the existence of W_s and of a second invariant manifold W_{cu}, called the center-unstable manifold. In this manifold, there could, in principle, be a two-dimensional submanifold which attracts solutions, but not at an exponential rate. For the particular case of the SH equation, this will not occur. We have argued – but not proved – in Section 19.1 that the two eigenvalues 1 correspond to *symmetries* of the problem, cf. (19.22). These symmetries are the translation invariance and the simultaneous change of ε and ω (keeping α fixed). Therefore, the submanifold corresponding to the indifferent eigenvalues is a manifold on which $\mathbf{N}(X) = X$. It follows that *the only solutions tending to zero are those corresponding to W_s.*

Since $\mathbf{N}$ only describes the discrete-time map, it is useful to reformulate the theorem for the Swift-Hohenberg equation itself. We parametrize the stable manifold by a parameter s. We now describe the *full* solution, not just the deviation from the stationary solution as in Theorem 21.1.

Theorem 21.2. *There is an $s_0 > 0$ and an $\varepsilon_0 > 0$ such that for every s satisfying $0 < s < s_0$ and every ε satisfying $0 < \varepsilon < \varepsilon_0$ and every ω satisfying*

$$3\varepsilon^2 > 2(1-\omega^2)^2 , \tag{21.12}$$

the following is true: There is a function

$$v_{s,\varepsilon,\omega} : \mathbf{R}^+ \to \mathbf{R}^4 ,$$

such that $\|v_{s,\varepsilon,\omega}(0) - u_{\varepsilon,\omega}(0)\| = s$, *and*

$$\big(\alpha - (1+\partial_x^2)^2\big)v_{s,\varepsilon,\omega}(x) - v_{s,\varepsilon,\omega}^3(x) = 0 .$$

Here, $u_{\varepsilon,\omega}$ *is a stationary solution of the SH equation with frequency* ω *and leading term* $2\varepsilon\cos(\omega x)$. *Furthermore, at* $s = 0$, *one has* $\partial_s v_{s,\varepsilon,\omega}(0) = \text{const.}\ v_0(0)$, *where* v_0 *is defined by (21.8). The constant* α *is determined by the relation (17.18).*

Remark. In order to keep the formulation of the theorem relatively short, we have omitted a uniformity condition. Instead of (21.12) one should really require

$$3\varepsilon^2 > \rho \cdot 2(1-\omega^2)^2 , \tag{21.13}$$

for some $0 < \rho < 1$, and the bounds on ε_0 and s_0 will then depend on ρ.

Remark. According to our spectral analysis, the function $v_{s,\varepsilon,\omega}(x)$ decays like $e^{-\kappa x}$ as $x \to \infty$. The constant κ is given by

$$\begin{aligned}\kappa &= \frac{\sqrt{3}\varepsilon}{2}\sqrt{2 - 4\frac{(1-\omega^2)^2}{3\varepsilon^2}} \\ &= \tfrac{1}{2}\sqrt{6\varepsilon^2 - 4(1-\omega^2)^2} \\ &= \sqrt{\frac{3}{2}\varepsilon^2 - (1-\omega^2)^2} .\end{aligned}$$

This point corresponds to the intersection point of the graph with the x-axis in Fig. 16.

21.4. Counting the Solutions

We now come to the interesting part of our study, namely exploiting the existence of solutions which decay exponentially. We want to glue them together to form a solution on the full real line.

We consider the SH equation with a *fixed parameter* $\alpha > 0$:

$$\big(\alpha - (1+\partial_x^2)^2\big)u(x) = u^3(x) .$$

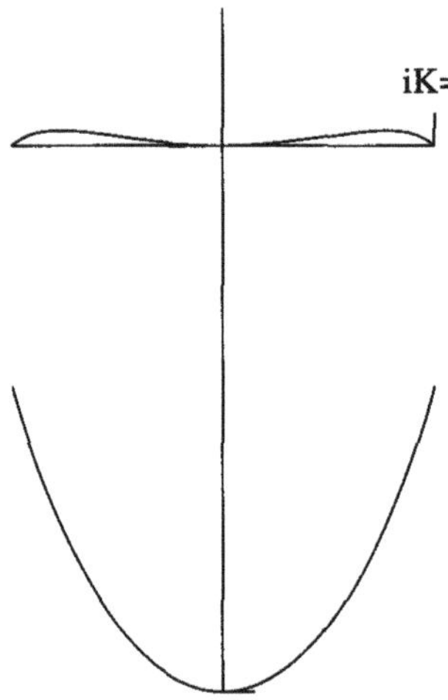

Fig. 16: The two eigenvalues for A_κ when κ is purely imaginary. We see that there are 4 values of κ where the eigenvalue is zero, namely $iK \approx \pm 0.75$ and a double eigenvalue at $iK = 0$. We use the scaled variables W, and K. The vertical axis represents the spectrum.

We restrict our attention to the positive half-line. We have first of all a one-parameter family of *stationary solutions*, parametrized, for example, by the amplitude ε. Then, ω is given by the relation

$$\lambda(\varepsilon) + (1 - \omega^2)^2 = \alpha ,$$

where λ is determined by (17.18) (implicitly, for that ω). Since we know from perturbation theory that $\lambda(\varepsilon) \approx 3\varepsilon^2$, the above equation has a unique solution. The existence of these stationary solutions is guaranteed in a parameter range described by Theorem 17.1. For sufficiently small α, the set of solutions is certainly not empty. If we consider now such a stationary solution u_ε, we can translate it in the x direction. We want to study the effect of this translation on the vector $X(u_\varepsilon, x_0)$, where $x_0 > 0$ is an arbitrary point on the real line. Recall that $X(v, x)$ was defined by

$$X(v, x) = \left(v(x), \partial_x v(x), \partial_x^2 v(x), \partial_x^3 v(x)\right) .$$

As we translate a solution in space, $(T_z v)(x) = v(x + z)$, and we let z vary in $\mathbf{R}$, the vector $X(T_z u_\varepsilon, x_0)$ draws a curve in $\mathbf{R}^4$ which is a topological circle, because the stationary solution is periodic. This curve is smooth, since we showed that the periodic solutions are analytic. For small ε it almost looks like a circle of radius 2ε, since the solution is essentially of the form $2\varepsilon\cos(\omega x)$. As we vary ε, and hence also ω, the set of these circles spans a two dimensional surface in $\mathbf{R}^4$. We see that we are in the situation of something like a Hopf bifurcation.

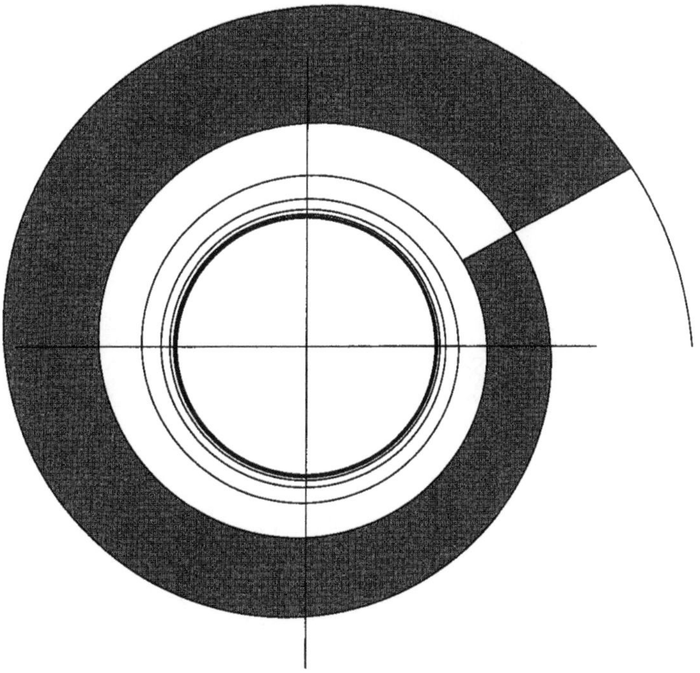

Fig. 17: A schematic representation of the available parameters and their symmetries, for fixed ω. The circle represents periodic solutions and the radius is about 2ε. Going along the circle corresponds to translations (that is, the orbit of the vector $X(T_z u_\varepsilon, x)$ described in the text). The radial direction represents the parameter t. The shaded region shows the set of solutions as one considers one set of t's and one period of translations. The full parameter set is the whole spiral, plus another spiral approaching the circle from the inside.

We now assume that, in addition to the restrictions on ε and ω already spelled out above, ε and ω satisfy the assumptions of Theorem 21.2, and, more precisely, the equation (21.13). There is a one parameter family of such values. According to Theorem 21.2, there is now a one-parameter family of solutions $v_{\varepsilon,t}$, decaying exponentially towards u_ε as $x \to \infty$. That is, $|v_{\varepsilon,t}(x) - u_\varepsilon(x)| \to 0$, and the same is true for the first 3 derivatives (and hence for all derivatives). These solutions can be translated with respect to the laboratory frame. In this light, the exponentially decaying solutions lead, upon translation in the x direction to curves $T_z X(v_{\varepsilon,t}, x_0) \in \mathbf{R}^4$ which spiral towards the periodic solutions. There is a one-parameter family of such curves for every choice of ε within the above limits, and these solutions are parametrized by the parameter t of the preceding section. Therefore, there are *three free parameters* for the solution of the SH equation on a half-space:

–The bifurcation parameter ε,
–the translation x,
–and the "amplitude" t.

These parameters vary in an open set which looks like an annulus. A picture of this domain, for fixed α, is shown in Fig. 18.

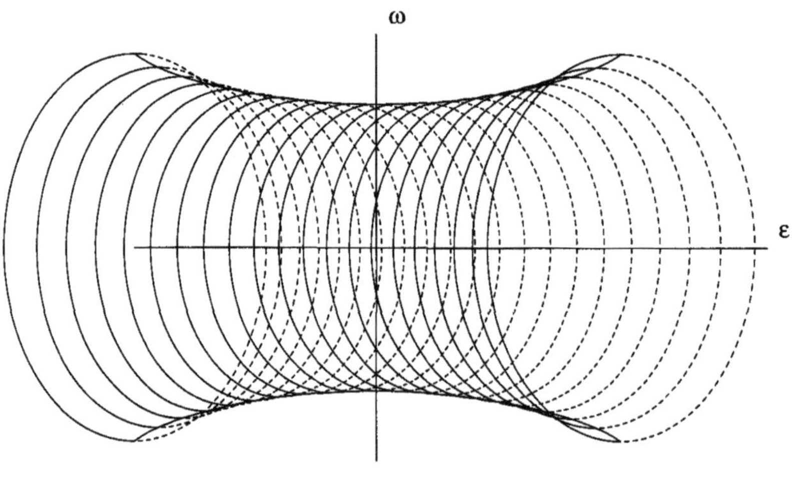

Fig. 18: A three-dimensional representation of the set of stationary solutions for the SH equation with fixed α. The horizontal axis is the bifurcation parameter ε and the vertical axis is the frequency. Going along the circle corresponds to following $X(T_z u_\varepsilon, 0) \in \mathbf{R}^4$ as z varies in $\mathbf{R}$.

We next discuss the glueing process. To make the exposition sufficiently simple, we will stick to a concrete example, but the generalization should be obvious. For the time being, we only address the question of differential equations whose coefficients are constant outside a compact domain on the real line. We consider a *fixed* equation and we ask for the family of solutions, for "arbitrary" ω. The example is very simple: Let $A(x)$ be a function which is constant for $|x| > 1$, and satisfies

$$A(x) = \begin{cases} \alpha_L, & \text{if } x \leq -1 \\ \alpha_R, & \text{if } x \geq 1 \\ \alpha(x), & \text{if } |x| < 1 \end{cases} ,$$

where $\alpha(x)$ interpolates smoothly between α_L and α_R. Furthermore, we assume $A(x)$ is strictly positive. We define a function $B(x)$ in a similar way, depending on β_L, β_R, and $\beta(x)$. Consider now the equation

$$\big(A(x) - (1 + \partial_x^2)^2\big)u(x) - B(x)u^3(x) = 0 . \tag{21.14}$$

For $x > 1$ and for $x < -1$, we find, if the parameters α_i and β_i are suitably chosen, two three-parameter families of solutions, as described above. The solution for $x > 1$ can be smoothly continued from $x = 1$ to $x = -1$ by integrating the differential equation (21.14). Since the differential equation is of fourth order, we need 4 of the 6 parameters to match the solution from the right with that from the left. This matching will always be possible if the parameters do not happen to be accidentally degenerate. In that case we remain with a *two parameter family of stationary solutions* to the equation (21.14). When the functions A and B are constant, we are (up to a trivial scaling) in the case of the SH equation. Then, the two free parameters are the "amplitude" ε and space translation. They determine ω. In the case of a true inhomogeneity, the above argument shows that there is still a two parameter family of solutions, but there are 6 complicated nonlinear relations between the quantities ε_L, ε_R, ω_L, ω_R, t_L and t_R, as well as the two translations T_{xL} and T_{xR} relative to the frame in which the perturbation is G. The solvability of these relations depends on a transversality condition for these relations. Since there is no "physical" degeneracy left in the problem, one can hope that for a generic perturbation the problem really has a two parameter family of solutions. A verification seems only feasible either in perturbation theory, or by numerical calculation for a concrete problem. It should be noted that all the steps of the solution in half-spaces follow from theorems which are constructively implementable, such as the contraction mapping theorem. Therefore, numerical calculations should work very well.

CHAPTER VI
MULTISCALE ANALYSIS

22. Rescaling

In Chapters I to V we have encountered a multitude of examples in which two phenomena were intimately connected:

- Destabilization, and
- Rescaling.

We want to argue now that this connection is not accidental, and we shall see in the next section and in Chapter VII an instance where 2 different scalings occur, leading to a so-called multiscale analysis.

The destabilization is always associated with a critical point in the underlying system. We have shown in particular that zero eigenvalues (or purely imaginary eigenvalues) are responsible for this destabilization. From the purely dimensional point of view, the rescaling becomes necessary because, when the real part of an eigenvalue vanishes, a length scale is lost and has to be re-fixed. This is also a well-known problem in Statistical Mechanics, where the correlation length is infinite at the critical temperature.

Let us informally summarize the problems encountered so far to make this point clear.

1. Bifurcation from a point attractor. We consider a vector field depending on a parameter. As the parameter is changed and an eigenvalue goes through zero, three quantities are rescaled: the parameter (deviation from the bifurcation point), the amplitude of the destabilized mode (deviation from the zero amplitude), and the time (critical slowing down). Note that for the Hopf bifurcation, the time of rotation is not rescaled. In Section 10 we have seen in (10.1) that the problem

$$\frac{dx}{dt} = -\lambda - x^2(t)$$

is best reparametrized by changing to the variables in which the amplitude x is the parameter. We have also seen that the time should be rescaled, leading to

$$x_\varepsilon(t) = \varepsilon y_\varepsilon(\varepsilon t) \ ,$$
$$\lambda_\varepsilon = -\varepsilon^2(1 + \mu_\varepsilon) \ ,$$

and the related regular equation

$$\frac{dy_\varepsilon(z)}{dz} = (1 + \mu_\varepsilon) - y_\varepsilon^2(z) \ .$$

2. The Crandall-Rabinowitz Theorem (Theorem 14.1).
Again, we recall that the amplitude is the relevant parameter, and the rescaling of the amplitude in terms of the bifurcation parameter leads to a regular operator (14.1).

3. Fronts for the SH equation. We shall see in Chapter VII that when the periodic stationary solution destabilizes in space, there will be a second scaling (in addition to the amplitude) with respect to the time. The limiting equation is then the real amplitude equation.

23. Multiscale Analysis with Continuous Spectrum

The preceding section gave an account of several problems where a multiscale analysis is adequate. We shall work out in detail the example of the Swift-Hohenberg equation in Section 28. Here, we put into perspective this more ambitious enterprise, which is really multiscale analysis for problems with continuous spectrum. While the approach in this section is mathematically incomplete, it will explain in an intuitive fashion how the new scales come about.

We have explained in Section 11 how to analyze bifurcations from a point attractor using perturbation theory for dynamical systems with discrete spectrum. We now extend these ideas to situations with continuous spectrum. We want to emphasize first that this theory is still in its infancy (some will say it is not even born). However, even though there are only very few rigorous results, the method is used very frequently and successfully in the physics literature. We shall follow as much as possible the lines of reasoning of Section 11 mentioning also some open questions.

We begin by explaining some ideas on multiscale analysis in a time-dependent situation. We consider the SH equation, close to the bifurcation point. This equation is

$$\partial_t W = \big(\alpha - (1 + \partial_x^2)^2\big) W - W^3 \ . \tag{23.1}$$

We have seen that the stationary solutions for small $\alpha > 0$ are very close to the function $u_0(x) = 2\varepsilon\cos(x)$. Here, we are taking $\omega = 1$ and this means that $\alpha \approx 3\varepsilon^2$. We are now interested in time-dependent solutions near u_0, and we search them in the form

$$W(x,t) = \varepsilon A(\varepsilon x, \varepsilon^2 t)e^{ix} + \varepsilon\bar{A}(\varepsilon x, \varepsilon^2 t)e^{-ix} .$$

A possible equation for A – it is not unique – is then

$$\begin{aligned}\partial_\tau A(y,\tau) &= 3A(y,\tau) + (4\partial_y^2 - 4i\varepsilon\partial_y^3 - \varepsilon^2\partial_y^4)A(y,\tau) \\ &\quad - A(y,\tau)|A(y,\tau)|^2 \\ &\quad - 3A^3(y,\tau)e^{2iy/\varepsilon} .\end{aligned} \tag{23.2}$$

This equation has a high order differential operator with small coefficient and an oscillatory term. When we set $\varepsilon = 0$ and neglect the oscillatory term, (23.2) reduces, up to a rescaling, to the complex amplitude equation

$$\partial_\tau A_0(y,\tau) = 3A_0(y,\tau) + 4\partial_y^2 A_0(y,\tau) - A_0(y,\tau)|A_0(y,\tau)|^2 . \tag{23.3}$$

We now want to make precise the sense in which the solutions for (23.2) are close to those of (23.3).

Conjecture 23.1. *Consider a function $v_0(x) : \mathbf{R} \to \mathbf{C}$, with v_0 and $\partial_x v_0$ in L^∞. Let $\tau_0 > 0$ be given. For sufficiently small ε and all $\tau < \tau_0$, the solutions of (23.2) and (23.3) with initial condition v_0 stay close (within $\mathcal{O}(\varepsilon)$) to each other in L^∞.*

This means that the solutions W_ε of the original problem, (23.1), for $\alpha = 3\varepsilon^2$ stay, for a time of order ε^{-2}, within $\mathcal{O}(\varepsilon^2)$ in L^∞ of

$$\varepsilon A_0(\varepsilon x, \varepsilon^2 t)e^{ix} + \varepsilon\bar{A}_0(\varepsilon x, \varepsilon^2 t)e^{-ix} ,$$

where A_0 solves (23.3).

Sketch of argument. To simplify the discussion, we consider instead of (23.2) the simpler equation

$$\partial_\tau w(y,\tau) = (\partial_y^2 - 1 - \varepsilon^2\partial_y^4)w(y,\tau) - (1 + s\cos(y/\varepsilon))\cdot w^3(y,\tau) , \tag{23.4}$$

and we compare it to the same equation with $\varepsilon = 0$ and $s = 0$. Consider first the simpler problem when $s = 0$, but $\varepsilon \neq 0$. In this case, one could first be inclined to believe that the term $\varepsilon^2\partial_y^4$ is totally irrelevant. This is not true, and for example, the sign of this term is crucial. This can be seen best for the Fourier transform

$$e^{-t(\xi^2+1+\varepsilon^2\xi^4)}$$

of the linearized evolution operator. Therefore, such things as inverses and resolvents of the differential operator have to be studied by spectral theory, and not by ordinary perturbation theory. We will develop some of these ideas in Section 27 and in Section 28.2.

As a second partial problem, we consider (23.4) without the fourth order term, but with $\varepsilon \neq 0$ and $s = \mathcal{O}(1)$. In this case, the properties of the differential operator are extremely well-known, since it is the heat kernel. To simplify the discussion even further, we consider a model problem

$$\partial_\tau w(y,\tau) = \partial_y^2 w(y,\tau) + e^{iy/\varepsilon} f(y,\tau) , \tag{23.5}$$

where f is a given function which, together with its derivative, is in L^∞. This problem can be solved explicitly in terms of the heat kernel. For zero initial data, one gets

$$w(y,\tau) = \text{const.} \int_0^\tau dt \int dx \, \frac{e^{-\frac{(y-x)^2}{4(\tau-t)}}}{\sqrt{\tau-t}} e^{ix/\varepsilon} f(x,t) .$$

Now one integrates by parts using

$$e^{ix/\varepsilon} = -i\varepsilon \partial_x e^{ix/\varepsilon} .$$

Since we have assumed that f and its derivative (with respect to x) are bounded, the term in which the derivative acts on f poses no problem. The other term is bounded by an expression of the form

$$\text{const.}\ \varepsilon \int_0^\tau dt \int dx \frac{|y-x|}{(\tau-t)^{3/2}} e^{-\frac{(y-x)^2}{4(\tau-t)}} .$$

This expression is of order ε. We expect the proof of the conjecture to be a combination of the two arguments given above. Since the difference between the solutions is small, the third power nonlinearity can be reduced to inhomogeneities, linear terms, plus small terms, by expanding

$$(w+\delta w)^3 = w^3 + 3\delta w \cdot w^2 + \mathcal{O}((\delta w)^2) .$$

The above method can be developed to provide a systematic multiscale perturbation theory to all orders in the bifurcation parameter ε.

We have not discussed the complete formal solvability of the expansion. In the above case this is easy to show. However, the computations become rapidly very involved and can be best done with computer algebra. Maslov [M3] presents some examples where even the formal expansion is nontrivial, because the transport equation, which is the analogue of (23.3), may not have a solution.

We also mention that in some cases one can shorten the computation by using special symmetry properties of the equation, see for example [CF] for an extensive use of invariance properties.

More complicated nonlinear effects have to be taken into account *if the bifurcation occurs at zero Fourier mode. A typical example for* this phenomenon is the limit $h \to 0$ in the Schrödinger equation. In an equation such as (23.2), but without the factor $\exp(2iy/\varepsilon)$, one can get still get stationary solutions such as those discussed in Section 11. But now there will also be some more interesting nonlinear WKB phenomena, that is, rapidly oscillating solutions, with "frequency" of order $1/\varepsilon$. See Maslov[M2] for some examples.

A famous multiscale analysis in the context of hydrodynamical equations is the analysis of the Rayleigh-Bénard problem by Newell, Whitehead, and Segur; see, for example [N]. We refer to [Man] for a discussion of this example. Another interesting example is provided in the paper of Zippelius and Siggia [ZS]. They show, for a hydrodynamic equation, how perturbation theory is applied. In particular, they find that in leading order, not only terms with fast and slow variables appear, but also a term which contains only the slow variables. This is interpreted as a sign of vorticity.

Many other similar computations exist in the literature, leading to amplitude equations. Another well-known example is the Kuramoto-Sivashinsky equation which occurs in problems which have a phase invariance and lead to interesting problems of phase dynamics and phase turbulence.

There are a few instances where the computation has been done rigorously, that is, where it has been shown that the formal power series corresponds to a true solution of the initial equation. We refer to [Ki], [IMD] for such a mathematical discussion. In this book, we explain most of another such example [CE1], in Section 28.

CHAPTER VII
FRONTS

If a physical system has several equilibrium states, or, as we have seen in earlier sections, several "quasistationary solutions," then we can ask the following question: What happens in an (infinite) system which is one equilibrium state near to infinity on "one side" and in another state on the other side? There will be an "interface" between the two phases and we may then ask what the motion of this interface is. If, intuitively, one phase "wins" over the other by invading its region, then we shall say that a front propagates into that region. In general, a stable solution wins over an unstable one. This definition excludes solitons, which are solutions connecting two stable stationary solutions. Fronts appear in many domains of science and it is of great importance to understand their nature. There are many different types of fronts: solitons, shocks, dissipative fronts. In this book, we only study fronts of the last type.

It should be pointed out that there is an extremely interesting subject which we do not cover here, namely the question of spinodal decomposition. This problem describes the situation where the system is in one of the two phases in several disconnected regions of space, and the two phases are approximately in equilibrium with respect to each other. They may then shrink and grow due to fluctuations in the supporting medium, for example, thermal fluctuations of a liquid in which the two phases coexist. The question is which "wins" in the long run, and if so, how the phase which takes over tends to fill up the space occupied by the other phase. Except for very simple models – such as the Ising model in Statistical Mechanics – it is already a nontrivial problem to associate phases at different points in space to a given configuration. A description of some results can be found in [GD].

It should also be noted that in space-dimension $d > 1$ there is a difficulty because an interface between two phases can fold back on itself. This kind of free boundary problem is only understood in very special situations. It has also been shown that singularities can occur in finite time. An example of this phenomenon is given in [SB].

Fronts have been studied in many contexts. In this book, we concentrate on fronts connecting an unstable state to a stable state, or to a marginally stable state. In this situation saddle-node connections seem to be the most adequate tool for study, while other types of fronts need a different set of mathematical tools [Fi].

24. Hamiltonian Formalism for Second Order Equations

Equations of the type

$$\partial_t u(x,t) = \partial_x^2 u(x,t) - F(u(x,t)) , \tag{24.1}$$

where x varies in **R**, can be discussed in a simple manner, using a Hamiltonian formalism. A typical choice for F is $F(q) = q^3 - q$. The following discussion basically only works for second order differential equations. In a moving frame, we want to consider solutions of the form $u(x,t) = v(x - ct)$. Then the equation for v is

$$\partial_z^2 v(z) + c\partial_z v(z) - F(v(z)) = 0 . \tag{24.2}$$

We now consider z as the *time* in a dynamical system, with $v(z_0)$ and $\partial_z v(z_0)$ as the initial conditions. Then the equations for $p(t) = \partial_z v(z_0+t)$ and $q(t) = v(z_0 + t)$ are

$$\begin{aligned} \partial_t q(t) &= p(t) , \\ \partial_t p(t) &= -cp(t) + F(q(t)) . \end{aligned} \tag{24.3}$$

If we define $V(q) = -\int_0^q F(s)ds$, then we see that (24.3) are the Hamiltonian equations for a system evolving in a potential V, and that c is a *friction term* in this evolution. We can now reformulate the front existence problem for the equation (24.1) as a problem for the Hamiltonian system (24.3). Clearly, the points q_0 where $F(q_0) = 0$ give rise to stationary solutions $u(x,t) = q_0$ of (24.1). These solutions correspond to equilibrium points of the associated Hamiltonian problem. The stationary solutions will be called stable if the corresponding Hamiltonian equilibrium point is stable, and unstable otherwise. If the equilibrium point is a minimum of the potential V, the equilibrium is stable. This will be the case when $\partial_q F(q_0) > 0$, and instability will correspond to $\partial_q F(q_0) < 0$. (The case of zero derivative is more delicate.) As we have said before, a *front* is a solution connecting a stable stationary solution to an unstable one, that is, in the Hamiltonian system it is a motion leading (in infinite time) from an unstable equilibrium point to a stable equilibrium point. Note that the above analysis says nothing about the *global* phase portrait.

24.1. The Real Amplitude Equation

To simplify the discussion, we restrict our attention to the case $F(q) = q^3 - q$. The potential V equals $-q^4/4 + q^2/2$, so that $q = 0$ is a stable equilibrium, and the $q = \pm 1$ are unstable equilibria. Initial conditions with $p = 0$ and $|q| > 1$ escape to infinity. These, and other initial conditions escaping to infinity (motions going outside the interval $[-1, 1]$) will not be considered here. A soliton is, in this language, a motion of a mass point, which starts at the unstable extremum $q = +1$ and reaches the unstable extremum $q = -1$ after an infinite time. Solitons exist in this case only for $c = 0$. This can be seen from the Hamiltonian picture: A particle starting at height $V(1) = 1/4$ cannot reach again height $1/4$ if there is friction.

In the Hamiltonian formalism, we may now ask: What are the c for which (24.2) has *positive* front solutions satisfying $v(-\infty) = 1$, $v(+\infty) = 0$? This corresponds to a motion in which a mass point "falls down" from a unstable equilibrium point and reaches a stable equilibrium point without "overshooting." It is intuitively clear that there must be a smallest damping c for which this situation occurs. This is called "critical damping." For every smaller damping, the solution will overshoot; every larger damping is stronger and the particle will creep into the equilibrium position. It so happens that for the potential at hand, that is, for $F(q) = q^3 - q$, critical damping occurs at $c = 2$.

We now turn to analysis of the system (24.3) as a dynamical system. It is of the form

$$\partial_t y(t) = X(y(t)) ,$$

where X is a vector field whose two components $y = (p, q)$ are given by (24.3). This vector field has three critical points (points where X is zero), namely $p = 0$, $q = 0$ and $p = 0$, $q = \pm 1$. The derivative of X at these fixed points is the matrix

$$\begin{pmatrix} -c & -1 + 3q^2 \\ 1 & 0 \end{pmatrix} ,$$

whose eigenvalues λ are

$$\lambda = \frac{-c \pm \sqrt{c^2 - 4}}{2} \quad \text{when } q = 0 ,$$

$$\lambda = \frac{-c \pm \sqrt{c^2 + 8}}{2} \quad \text{when } q = \pm 1 .$$

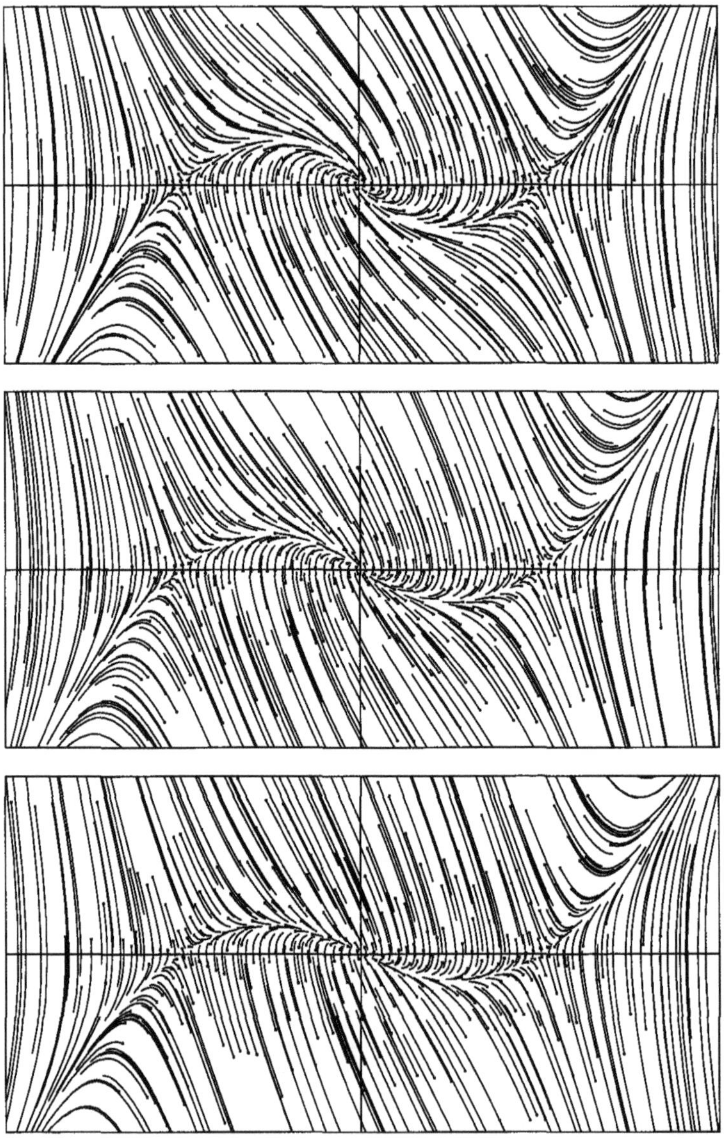

Fig. 19: Three phase portraits for the Hamiltonian vector field for the real amplitude equation in a frame moving with speed $c = 1.5, 2, 2.5$. The horizontal axis is q, the vertical is p. Note the qualitative change from "overshooting" to "overcritical damping" as c increases. The flow lines start at the dots.

The eigenvalues at $|q| = 1$ are always of opposite sign. At $q = 0$ they are both negative when $c \geq 2$, they coincide when $c = 2$ and they are complex conjugate with negative real part when $0 < c < 2$. In Fig. 19, we show the phase portrait for the three cases.* (Note that the eigenvectors have the two components $(\lambda, 1)$.)

Let us consider the phase portrait in some more detail. At the fixed point $q = 1$, $p = 0$, we have, according to the stability analysis above, one stable and one unstable direction. Therefore, the only solutions of the original equation which reach the value 1 at $z = -\infty$ are those which are tangent to the unstable direction at $q = 1$, $p = 0$. Since the eigenvalues are of different size, there can be only one curve tangent to this direction. This curve leaves that point and will eventually reach the point $q = 0$, $p = 0$. Let us now assume $c > 2$. Then the two eigenvalues are distinct, and negative, and unless there is an accidental situation, the curve will be "pulled in" by the weaker (less negative eigenvalue). Fig. 20 shows this clearly.

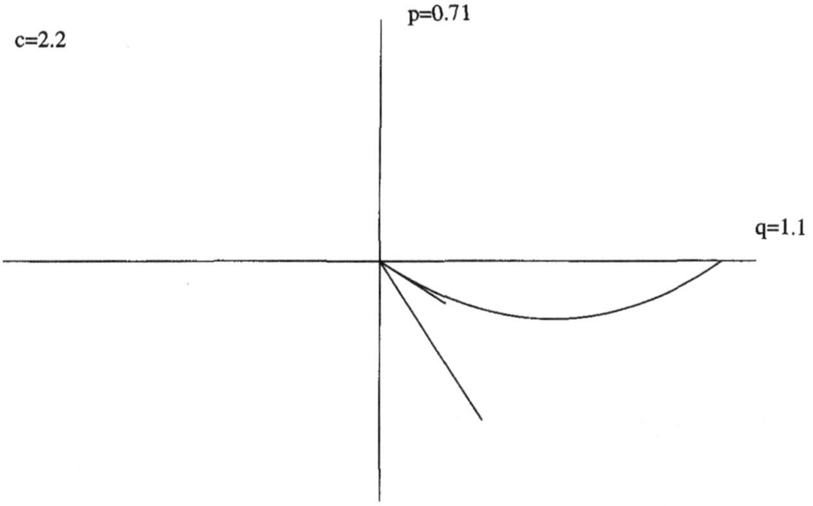

Fig. 20: The two straight lines indicate the directions of the two eigenvectors at 0. Their lengths are proportional to the eigenvalue. One sees clearly that the orbit starting at $p = 0$, $q = 1$ is "pulled in" by the weaker eigenvalue. The damping is overcritical.

* We thank G. Wanner for providing us with the programs to do these drawings [HNW].

Thus, we see that *the exponential ahead of the front decays with the slower of two possible decay rates.* Note also that the front solution corresponds to an orbit connecting two fixed points of the vector field. Such orbits are saddle connections.

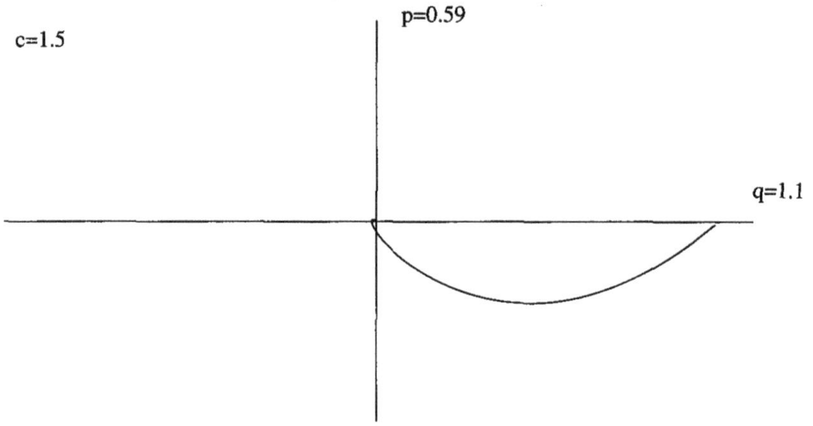

Fig. 21: An orbit for subcritical damping, shown in phase space. The curve "spirals in" (oscillates into equilibrium). The front is not a positive function in this case.

When $c = 2$ the two rates coincide. Next, when $0 < c < 2$, then there is an oscillatory motion, described in Fig. 21 in the phase space. In Fig. 22, we show the three cases as amplitudes.

24.2. The Complex Amplitude Equation

We next study the complex equation

$$\partial_t u(x,t) = \partial_x^2 u(x,t) + u(x,t) \cdot (1 - |u(x,t)|^2) . \tag{24.4}$$

The corresponding Hamiltonian is

$$H = \frac{|u'|^2}{2} + \frac{|u|^2}{2} - \frac{|u|^4}{4} .$$

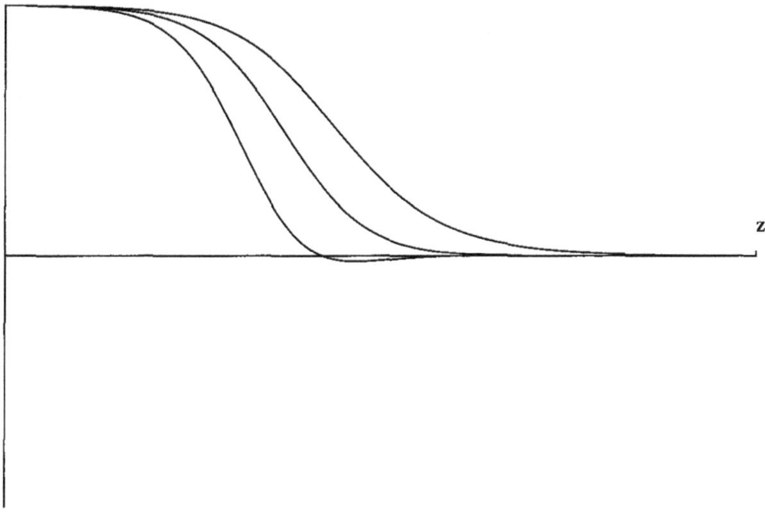

Fig. 22: Three typical fronts. The horizontal axis is the coordinate z and the vertical axis is the order parameter $u(z)$. The three curves correspond to $c < 2$, $c = 2$, $c > 2$ (lowest to highest curve).

Again, we are only interested in the local basin of attraction near $u = 0$, $u' = 0$. The phase portrait of this equation is somewhat more interesting than that of (24.1) and we shall get a richer set of possibilities for the fronts. We are interested in "modulated moving fronts," that is, solutions of the form

$$u(x,t) = e^{ikx}v(x-ct)e^{i\phi(x-ct)} , \tag{24.5}$$

where k, v, and ϕ are real. Assume that ϕ and v have limits as $x - ct \to \infty$. Then the form (24.5) means that we are looking for solutions which leave a periodic pattern – namely, e^{ikx} – in the laboratory frame. In a frame moving with speed c the corresponding solution has a complex amplitude which is similar to the one found for the equation (24.2). It is interesting to see that the decay of this amplitude is not totally trivial.

To study this last question, we re-express (24.4) in the new variables. The corresponding equations are (with primes denoting derivatives with respect to z),

$$0 = v''(z) + cv'(z) + v(z)(1 - (k + \phi'(z))^2) - v(z)^3 ,$$
$$0 = \phi''(z)v(z) + 2(\phi'(z) + k)v'(z) + c\phi'(z)v(z) .$$

Define now the quantities

$$\kappa(z) = \phi'(z) + k ,$$
$$q(z) = v'(z)/v(z) .$$

In terms of these quantities, the above system is transformed to

$$\begin{aligned} v'(z) &= q(z)v(z) , \\ \kappa'(z) &= -\kappa(z)(2q(z) + c) + ck , \\ q'(z) &= -q^2(z) - cq(z) - 1 + \kappa^2(z) + v(z)^2 . \end{aligned} \tag{24.6}$$

The only real fixed points of this system are, for $|k| \leq 1$, $c \geq 2$,

1) $$q = 0 , \quad \kappa = k , \quad v = \sqrt{1 - k^2} ,$$

2) $$q = -\frac{c}{2} + \tfrac{1}{2}\sqrt{c^2 - 4(1 - \kappa^2)} ,$$
$$\kappa = \sqrt{\tfrac{1}{2}(1 - \frac{c^2}{4}) + \tfrac{1}{2}\sqrt{(\frac{c^2}{4} - 1)^2 + k^2c^2}} , \quad v = 0 ,$$

3) $$q = -\frac{c}{2} - \tfrac{1}{2}\sqrt{c^2 - 4(1 - \kappa^2)} ,$$
$$\kappa = -\sqrt{\tfrac{1}{2}(1 - \frac{c^2}{4}) + \tfrac{1}{2}\sqrt{(\frac{c^2}{4} - 1)^2 + k^2c^2}} , \quad v = 0 .$$

The interpretation of these fixed points is as follows: The fixed point 1) corresponds to the stationary solution

$$u(x, t) = \sqrt{1 - k^2}e^{ikx} .$$

This solution is stationary in the laboratory frame. The solutions 2) and 3) correspond to zero solutions, but with different asymptotic behavior. We will construct, as in the case of (24.1), a modulated front by finding a saddle-connection between the solutions 1) and 2).

In the basis v, κ, q, the tangent map to the vector field in (24.6) is given by

$$\begin{pmatrix} q & 0 & v \\ 0 & -2q-c & -2\kappa \\ 2v & 2\kappa & -2q-c \end{pmatrix} .$$

Therefore, a simple calculation shows that there are one unstable and two stable eigenvalues at the fixed point 1). At the fixed point 2), the eigenvalues are

$$\begin{aligned} \lambda_1 &= q , \\ \lambda_{2,3} &= -c-2q \pm 2i\kappa . \end{aligned}$$

Thus, we see that all eigenvalues have negative real part. Their relative strength depends on k. In the region where $\operatorname{Re}\lambda_{2,3} < \lambda_1 < 0$, almost all solutions tending to zero are asymptotically tangent to the invariant eigenplane $\{u=0\}$. This case occurs for $k^2 \geq 1/9 - 2c^2/81$, and thus in a c region which is the exterior of an ellipse. Consider now one of the solutions going into the fixed point 3). Then its behavior at infinity is given by the orbit in the κ-q plane. We now fix q, κ, v to be a solution of the equation (24.6), given by 3). Then there is an orbit near that fixed point which is essentially given by the linear map, that is,

$$\begin{aligned} v(\tau) &\approx \text{const. } e^{\lambda_1 \tau} , \\ \kappa(\tau) &\approx \text{const. } e^{\operatorname{Re}\lambda_2 \tau} \cos(\operatorname{Im}\lambda_2 \tau) , \\ q(\tau) &\approx \text{const. } e^{\operatorname{Re}\lambda_2 \tau} \sin(\operatorname{Im}\lambda_2 \tau) . \end{aligned}$$

This means that the function u is, near $x=\infty$, given by

$$u(x,t) = e^{ikct} e^{q(x-ct)} e^{i \int^{x-ct} \kappa(\tau) d\tau} .$$

Note that this front has frequency and an amplitude which are *modulated* as is visible in Fig. 23.

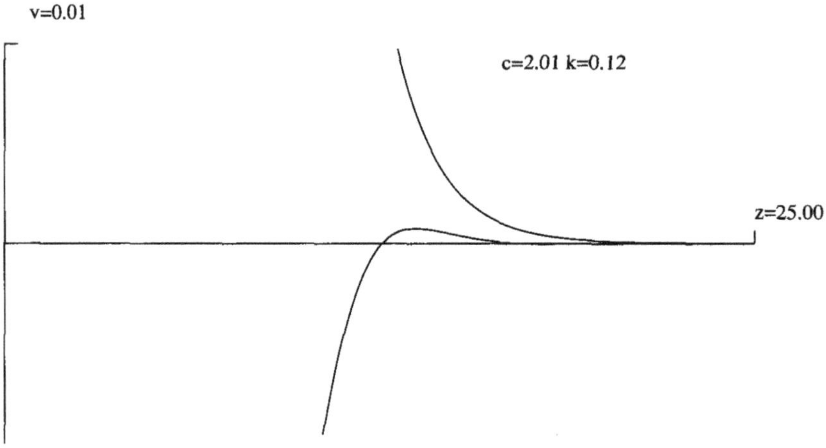

Fig. 23: The behavior of the solution ahead of the front for the complex amplitude equation. The upper curve is the amplitude v as a function of z. The lower curve is the derivative ϕ' of the phase. Observe that it is non-monotone.

25. The Maximum Principle and Comparison Theorems

In Section 24, we have seen that positive front solutions exist for the real amplitude equation for all $c \geq 2$. An intriguing question, raised by Langer and Müller-Krumbhaar [LMK], is to know which of the solutions is selected if we start from the unstable state. This question is not answered in full generality for arbitrary systems, but a beautiful solution has been given by Aronson and Weinberger for the case of semilinear parabolic equations (essentially in 1 dimension) [AW]. In this section, we explain their argument.

We consider differential equations of the form

$$\partial_t u(x,t) = \partial_x^2 u(x,t) + a(x,t)\partial_x u(x,t) + b(x,t,u(x,t)) \,. \qquad (25.1)$$

Such equations are called semilinear parabolic equations. For such equations, certain inequalities hold if the differential operator is of second order. These inequalities are useful in the proofs of velocity selection for the real amplitude equation. They provide a very elegant framework, but it must be stressed that no analogous results seem to be true for higher order differential equations.

We want to explain the basic idea for a special case of (25.1). We consider the differential equation

$$\partial_t u(x,t) = \partial_x^2 u(x,t) + a\partial_x u(x,t) + b(u(x,t)) . \qquad (25.2)$$

This is (25.1) with constant coefficients. Assume that we have two sets of initial data, $u_1(x,0)$ and $u_2(x,0)$, and suppose that $u_1(x,0) > u_2(x,0)$ for all $x \in \mathbf{R}$.
Then, for all $t > 0$ and all $x \in \mathbf{R}$, one has $u_1(x,t) > u_2(x,t)$.
This means that solution curves cannot cross.

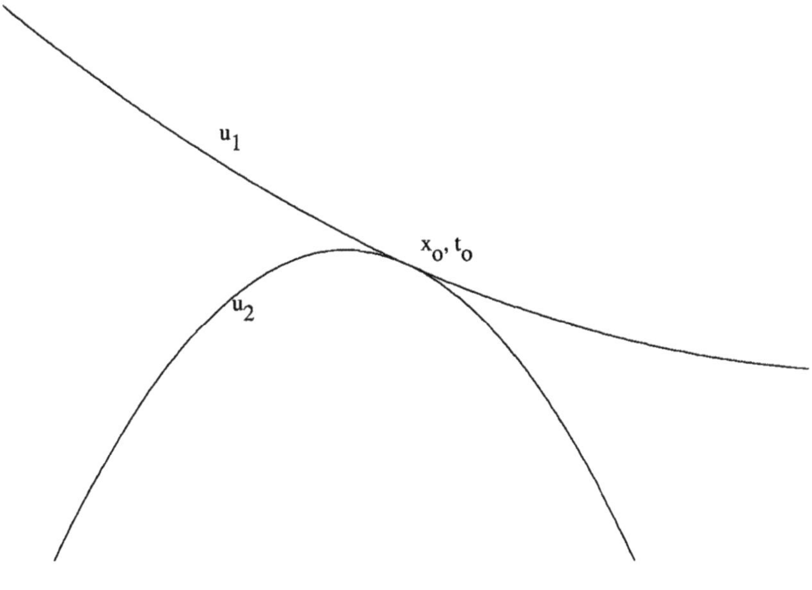

Fig. 24: The curve u_2 is in a position where it could cross u_1 at the point x_0, t_0. However, the curvature pushes it downwards.

We now give a plausible argument (not a proof!) of why this is true. Indeed, looking at Fig. 24, suppose that u_2 is about to cross u_1 at some point $x = x_0$ at the instant $t = t_0$. Then

$$u_1(x_0,t_0) = u_2(x_0,t_0) ,$$
$$\partial_x u_1(x_0,t_0) = \partial_x u_2(x_0,t_0) .$$

Since the curve u_2 is below u_1 we also have

$$\partial_x^2 u_2(x_0,t_0) < \partial_x^2 u_1(x_0,t_0) .$$

Finally, since the u_i coincide, we have $b(u_1(x_0,t_0)) = b(u_2(x_0,t_0))$. These equalities and (25.2) imply that

$$\partial_t(u_1 - u_2)(x_0,t_0) > 0 ,$$

that is, u_1 and u_2 get pushed apart, and hence refuse to cross.

Physicists are more familiar with the comparison theorem in a different guise: It says essentially that in a heat diffusion problem the temperature at *any* point of the sample and for any time is always lower than the maximal temperature at the initial time.

The non-crossing argument is basically sound, but is incorrect on 2 points:

- The second derivatives could coincide.
- The solutions could start crossing near $x = \pm\infty$.

The Comparison Theorem takes care of these problems. Before we state it, we note that the above conclusions would also hold if instead of the Equation (25.2), the functions u_1 and u_2 satisfy only inequalities, namely

$$\begin{aligned}\partial_t u_1(x,t) &\geq \partial_x^2 u_1(x,t) + a\partial_x u_1(x,t) + b(u_1(x,t)) ,\\ \partial_t u_2(x,t) &\leq \partial_x^2 u_2(x,t) + a\partial_x u_2(x,t) + b(u_2(x,t)) .\end{aligned}$$

Theorem 25.1. Comparison Theorem [F], [PW]. *Suppose that u_1 and u_2 are bounded continuous functions in $(x_1,x_2)\times[0,T]$, for some $T>0$. The quantities x_1 and x_2 can be finite or infinite. Assume the coefficient functions a, b are continuous and*

$$\begin{aligned}&\partial_t u_1(z) - \partial_x^2 u_1(z) - a(z)\partial_x u_1(z) - b(z,u_1(z))\\ &\geq \partial_t u_2(z) - \partial_x^2 u_2(z) - a(z)\partial_x u_2(z) - b(z,u_2(z)) ,\end{aligned} \tag{25.3}$$

for $z=(x,t)\in(x_1,x_2)\times(0,T]$ and

$$u_1(x,0) \geq u_2(x,0) , \qquad \text{for } x\in(x_1,x_2) . \tag{25.4}$$

If an x_i, $i=1,2$ is finite, then we also assume that

$$u_1(x_i,t) \geq u_2(x_i,t) , \qquad \text{for } t\in[0,T] . \tag{25.5}$$

Under these assumptions, one has, on all of $(x_1,x_2)\times[0,T]$, the inequality

$$u_1(x,t) \geq u_2(x,t). \tag{25.6}$$

Moreover, if u_1 and u_2 coincide in some point $(x_0,t_0)\in(x_1,x_2)\times(0,T]$ then they coincide for all $(x,t)\in(x_1,x_2)\times[0,t_0]$.

We have drawn this region in Fig. 25.

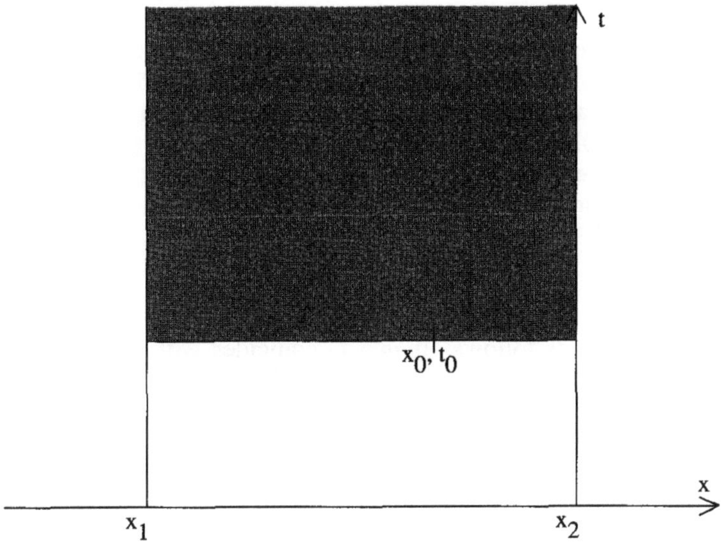

Fig. 25: An illustration of the "domain of influence" of the point x_0, t_0. The horizontal axis is space, the vertical is time.

Proof. For the proof, we refer to [F], [PW]. The original theorem is only formulated for b's which are independent of z.

Remark. As far as we can see, there is no derivation of this argument from the Hamiltonian picture of Section 24.

25.1. Some Applications

To understand the meaning of the Comparison Theorem it is best to give a few applications.

Example 1. Consider the real amplitude equation

$$\partial_t u(x,t) = \partial_x^2 u(x,t) - F(u)(x,t) , \qquad (25.7)$$

where $F(u) = -u + u^3$. Assume the initial data are $u(x,0) = u_0(x)$ and that $0 \le u_0(x) \le 1$, for all $x \in \mathbf{R}$. Then $u(x,t)$ satisfies for all $t > 0$ the inequality

$$0 \le u(x,t) \le 1 .$$

This is seen as follows: Take $u_1(x,t) \equiv 1$ and $u_2(x,t) = u(x,t)$ in the Comparison Theorem. Then (25.3) is an equality for $t > 0$, since both sides are zero (both functions are solutions of the equation). The inequality (25.4) holds by assumption. Therefore, (25.6) follows. The same argument applies when we compare the function $u_1 = u$ with the function $u_2 \equiv 0$. This example therefore shows that positive solutions must stay between zero and one. Note that if u_0 is not identically equal to 1 or 0, then one has

$$0 < u(x,t) < 1 \quad \text{for all } t > 0 .$$

This can be seen as follows: If $u(x,t)$ coincides with either u_1 or u_2 at some point (x,t), then by the Comparison Theorem it coincides for all x and all $t' < t$, and this contradicts the assumption that u is not identically 0 or 1.

Example2. The Hair-Trigger Effect. In this example we want to show that if $u_0(x) \in [0,1]$ and u_0 is not identically 0 then, for every x, one has

$$\lim_{t\to\infty} u(x,t) = 1 . \tag{25.8}$$

In other words, the *linear* instability of the zero solution implies a *nonlinear* instability of the solution. (This need not be the case in more general nonlinear problems; in particular, the condition $F'(0) < 0$ seems essential in the argument.) Fix some small positive time t_0. Since $u_0 \not\equiv 0$, we have by the last assertion of the Comparison Theorem that $u(x,t_0) > 0$. We define

$$\nu = \min\{u(x,t_0) \ : \ |x| < 2\frac{\pi}{|F'(0)|^{1/2}}\} .$$

Clearly, $\nu > 0$. Choose now an ε in $(0,\nu]$ and define the function $q_\varepsilon(x)$ as the solution of $q''(x) - F(q(x)) = 0$ with $q(0) = \varepsilon$ and $q'(0) = 0$. In view of the Hamiltonian formalism, this solution is obtained by "dropping" the particle at "height" ε in the potential $V(x) = -\int_0^x dy\, F(y)$, with initial speed 0. We consider next the time when this solution reaches the "height" 0. We denote by $\pm b_\varepsilon$ the two hitting times. For small ε, q satisfies $q''(x) \approx F'(0)q(x)$, that is, it behaves like a harmonic oscillator. The frequency is determined by $F'(0)$ (which we assumed to be different from zero), and hence we see that $b_\varepsilon \to \pi/|F'(0)|^{1/2}$ as $\varepsilon \to 0$. Thus, the hitting time is finite, as $\varepsilon \to 0$. Define now $v(x,0) = q_\varepsilon(x)$ when x is in the (central) connected region where q_ε is positive, and $v(x) = 0$ for all other x. This setup is described in Fig. 26.

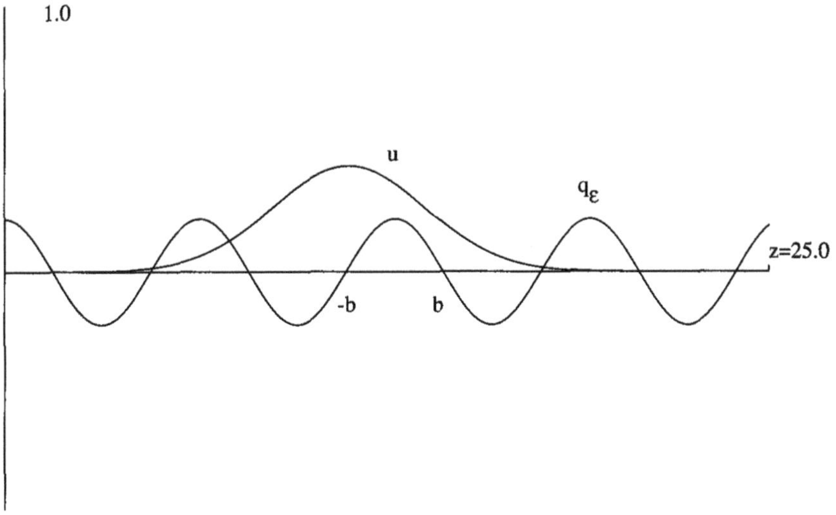

Fig. 26: The oscillating curve q_ε corresponds to an oscillatory motion about the equilibrium point. The function u lies above one of the oscillations, and will therefore be "pushed up" to the value 1.

Let now $v(x,t)$ be the solution with this initial condition. Using the Comparison Theorem, we have $v(x,t) \leq u(x,t)$ for all x, and $t > t_0$. If we can show that $v(x,t)$ tends to 1 the assertion (25.8) follows. The following lemma deals with this problem:

Lemma 25.2. *Assume $w(x,0)$ takes values in $[0,1]$ and has compact support. Assume furthermore that, on its support, $w(x,0)$ is a solution of the stationary equation. If w is the solution of (25.7) with initial data $w(x,0)$, then*

$$\lim_{t\to\infty} w(x,t) = \tau(x) ,$$

where τ is the smallest stationary solution of (25.7) which satisfies

$$\tau(x) \geq w(x,0) .$$

The convergence is uniform on compact sets of **R**.

We apply this lemma with $w(x,0) = v(x,0)$, where v has been defined above.

Proof. We only sketch the argument. By applying the Comparison Theorem to $w(x,t)$ and $w(x,t+\delta)$, one can show that $w(x,t)$ is a *non-decreasing* function of t. Since it is bounded above by 1, the limit exists. It is shown in [AW] that this limit is a stationary solution: Essentially, one shows that if the limit is not stationary it continues to grow.

Remark. Lemma 25.2 shows that no positive solution ever decreases in time! Since there are no stationary solutions between 0 and 1, we see that $u(x,t) \to 1$ as $t \to \infty$.

Example 3. In this example, and the following one, we repeat in a sense the analysis of Examples 1 and 2, but now in a *frame moving with speed c*. A stationary solution in a moving frame is a front in the laboratory frame. Thus the comparisons with stationary solutions we found in Examples 1 and 2 become comparisons with fronts. Also, the Equation (25.7) is replaced by

$$\partial_t u(x,t) = \partial_x^2 u(x,t) + c\partial_x u(x,t) - F(u)(x,t) , \qquad (25.9)$$

where $F(u) = -u + u^3$.

We look at initial conditions of the form shown in Fig. 27: they tend to 1 at $-\infty$ and to 0 at $+\infty$. To make the discussion simpler, we denote by U the solution of

$$\partial_t U(x,t) = \partial_x^2 U(x,t) + U(x,t) - U^3(x,t) ,$$

which moves with speed $c \geq 2$ (to the right), and is normalized by $U(0,0) = 1/2$. The existence of such a solution has been discussed in Section 24, where we showed that $U(x,t) = V(x-ct)$, with

$$V''(z) = -cV'(z) - V(z) + V^3(z) , \quad V(0) = 1/2.$$

Assume now that there are two constants t_1 and t_2 for which the initial data $u(x,0) = u_0(x)$ satisfy

$$V(x-t_1) \geq u_0(x) \geq V(x+t_2) .$$

Then the solution $u(x,t)$ will move *exactly with speed c*, since it is "sandwiched" between the solutions $U(x,t) = V(x-t_1-ct)$ and $U(x,t) = V(x+t_2-ct)$, cf. Fig. 27.

Remark. The analogue of Lemma 25.2 implies that *no small positive initial condition u_0 with compact support can propagate faster than the*

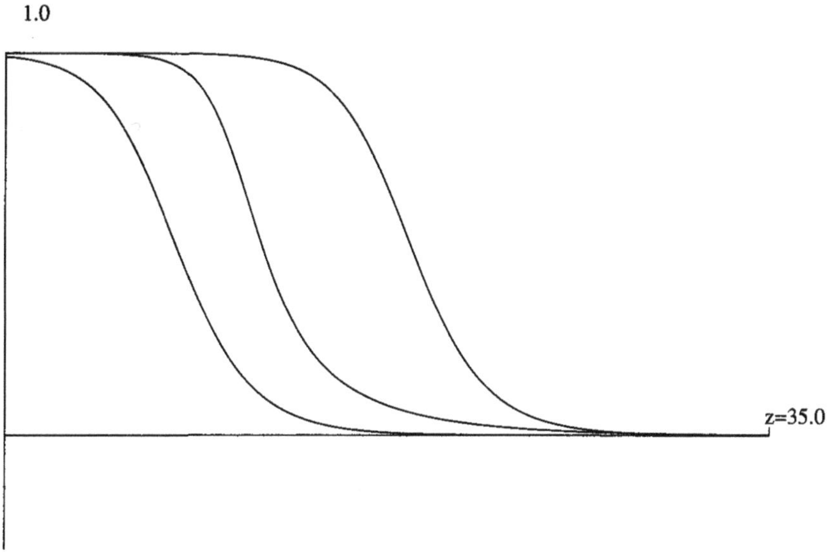

Fig. 27: Two fronts moving at the same speed to the right and an initial function u_0 which is trapped between the two fronts.

(positive) front with the slowest speed. To see this, we go to a frame moving with speed $c > 0$. We have seen in Section 24 that positive front solutions exist only if $c \geq 2$. These solutions are stationary in the frame moving with speed c. The Lemma 25.2 also holds for (25.9), as well, and says that in the frame moving with speed c the solution evolving from u_0 must stay below the "stationary" front solution. Since this holds for all $c > 0$ for which there is a positive front solution, the assertion follows.

Example 4. One can in fact do better than in Example 3: We now show that the speed is already determined by the behavior of the solution *ahead* of the front, that is, it is not necessary to be "sandwiched" to the left in Fig. 28.

Fix a $c \geq 2$, and let $\gamma_c = (c - \sqrt{c^2 - 4})/2$, that is, γ_c is the decay rate of a generic front moving with speed c, as shown in Section 24. Assume now that the initial data u_0 satisfy

$$0 < a_1 e^{-\gamma_c x} < u_0(x) < a_2 e^{-\gamma_c x} < 1 ,$$

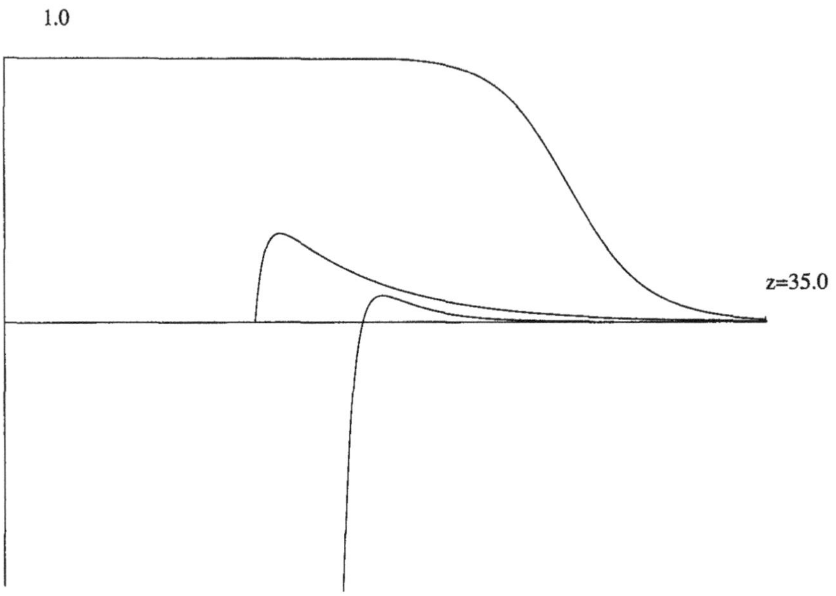

Fig. 28: An upper and a lower envelope moving to the right, and a function trapped between the two.

for $x > x_0$, and $u_0(x) \in [0, a]$ (with $a < 1$) for all x. At any time $t > 0$ the solution of (25.7) will be positive. Therefore, considering that time t as the "starting time," we may as well assume the $u_0(x) \in (0, 1)$, and that in addition it is bounded away from zero on compact sets in x. We now consider two comparison functions. We start with a lower function. It is constructed in analogy with Example 2. Choose any small "height" ε and define $q(x)$ as the solution of $\partial_x^2 q(x) + c\partial_x q(x) - F(q(x)) = 0$ with initial condition $q(0) = \varepsilon$, $q'(0) = 0$. This corresponds to dropping a particle in the potential $V(x) = -\int^x dy\, F(y)$, but now with friction c, whereas in Example 2, the friction was 0. For sufficiently small ε, q satisfies the equation

$$\partial_x^2 q(x) + c\partial_x q(x) \approx F'(0)q(x) .$$

If $c \geq 2$, the solution to this equation does not "overshoot," that is, it stays positive. To see this, we note that the solution does not overshoot – when $c \geq 2$ – if the point starts at height 1. Letting it start lower down at slower speed clearly gives it a lower speed at any height. Since the initial data satisfy $u_0(x) < a_2 e^{-\gamma_c x}$, we can always find an upper solution moving with the corresponding speed c, by translating the associated front sufficiently far to the right. It should be kept in mind that the above methods do *not* say what the exact motion of the growing front is. For that one needs more precise estimates.

Theorem 25.3. *Let $u(x,0)$ be a regular positive initial condition which converges to 1 for $x \to +\infty$, and satisfying $u(x,0) \sim \exp(-\gamma x)$ for large x (we refer to the original papers for a precise definition of this equivalence). Let c be defined by*

$$\begin{cases} \gamma_c = \gamma \text{ if } \gamma < 1\,, \\ c = 2 \text{ otherwise.} \end{cases}$$

Let ℓ_c denote a front solution moving with speed c. Then, if $u(x,t)$ is the solution at time t with initial condition $u(x,0)$ at time 0, there is a function $m(t)$ such that

$$\lim_{t \to \infty} m(t)/t = 0$$

and

$$\sup_{x \in \mathbf{R}} |u(x,t) - \ell_c(x - ct - m(t))| \xrightarrow[t \to \infty]{} 0$$

(see [Br] for logarithmic estimates on $m(t)$).

In other words, except for a shift, in the frame moving at the correct speed, the profile converges to the profile of the corresponding front. Moreover, if the decay of the initial condition is fast enough, the limiting speed is 2. We can say that 2 is the speed corresponding to the largest set of initial conditions.

This theorem can be extended to the more general nonlinearities mentioned above. If the nonlinearity is concave, the selected speed is always given by twice the square root of the slope of the nonlinearity at the origin. However, this may not be true for nonconcave functions. An extreme situation is given by a nonlinearity with slope zero at the origin and which is not identically zero. Then one can show that the selected speed is strictly positive. However, no explicit expression has been derived up to now for this speed.

26. The Stability Analysis for Fronts

The above theorem applies mostly to positive solutions. The reason is that the proof is based on the parabolic maximum principle. There is no reason to expect an extension of the proof to the CA equation, which is not a real equation, or to the SH equation which is a fourth order equation. It is therefore necessary to give another selection criterion which can be applied to more general situations. As we shall see, this can be done in terms of a local stability analysis of the fronts, but we shall lose the globality of the theorem of Aronson-Weinberger.

The more general selection criterion is known as the marginal stability criterion. It is a minimal set of stability properties that should be satisfied to have selection. Note, however, that this criterion may not be strong enough in all cases. The main idea is to perform a linear stability analysis for the fronts in the moving frame. One should first state precisely the space of admissible perturbations because the spectrum of the linearized evolution may depend strongly of this space. By analogy with the result of Aronson-Weinberger we shall consider a family of spaces $\mathcal{B}_\gamma$ defined by

$$\mathcal{B}_\gamma = \{h \mid h(x)(1 + e^{\gamma x}) \in \mathrm{L}^\infty(dx)\} .$$

This is a Banach space when equipped with the natural norm. As explained above, if we consider the stability of a front ℓ_c of speed c in the moving frame, this spectrum will depend of the value of γ determining the space $\mathcal{B}_\gamma$ of allowed perturbations. If γ is too large, the front will be stable, and similarly for small γ the front will be unstable (see [S]). A natural choice for γ is the value γ_c which gives the decay of the front. In $\mathcal{B}_{\gamma_c}$ we expect the spectrum to be at best marginally stable because of the mode corresponding to the translation of the front. In fact, since we want to include this mode for physical reasons, we impose the condition $\gamma \le \gamma_c$. The result of the spectral analysis can then be formulated as follows.

Proposition 26.1. *For the real amplitude equation, if $c < 2$ the front ℓ_c is unstable in $\mathcal{B}_{\gamma_c}$. If $c \ge 2$, the front ℓ_c is marginal in $\mathcal{B}_{\gamma_c}$.*

Note that the above statement is not precise: It in fact only deals with the position of the essential spectrum. There could be an infinite number of isolated eigenvalues in the right half-plane which accumulate at the continuous spectrum, and they are not captured by the Proposition 26.1.

Proof. This proof is an archetype of similar proofs we will use throughout. Consider first a differential operator $A = P(-i\partial_x)$, with constant coefficients. Then we have seen, by Fourier analysis, that the spectrum of A in L^2 and in L^∞ is equal to the set

$$\Sigma = \{P(\xi) \ : \ \xi \in \mathbf{R}\} \ .$$

Note that A has no discrete spectrum. Consider next the operator for the front problem (25.9) in the frame moving with speed c. It is

$$A = \partial_x^2 + c\partial_x - F'(\ell_c)(x) \ .$$

This operator does not have constant coefficients, since ℓ_c goes from 1 to 0. However, the coefficients *have limits* as $x \to \pm\infty$. This leads to the general question of the continuous spectrum of an operator whose coefficients have limits. The answer is given by

Proposition 26.2. *The essential spectrum of a real differential operator $A = P(x, -i\partial_x)$ whose coefficients have limits and whose highest derivative term has constant coefficient is the union of the essential spectra on each side:*

$$\Sigma = \{P(+\infty, \xi) \ : \ \xi \in \mathbf{R}\} \cup \{P(-\infty, \xi) \ : \ \xi \in \mathbf{R}\} \ .$$

Proof. For a proof, see Rota [Ro]. In the complex case the formulation is somewhat more complicated but the ideas remain the same; see [CE2]. Basically, the proofs are a rigorous version of the following strategy. "Cut" the operator into two pieces, left and right. Consider for example the left piece on $\mathrm{L}^\infty(-\infty, 0]$. This operator is equal to an operator with constant coefficients plus a remainder which is relatively compact with respect to the (constant) highest order part. Therefore, the continuous spectrum is given by the limiting operator. However, there can be a discrete additional spectrum. It corresponds to a bound state of the "quantum-mechanical problem"

$$-\partial_x^2\psi(x) - c\partial_x\psi(x) + F'(\ell_c)(x)\psi(x) = E\psi \ ,$$

with negative E. The above discussion is valid in $\mathrm{L}^\infty(dx)$. We are really interested in stability in the spaces $\mathcal{B}_{\gamma_c}$ and only ahead of the front. The essential spectrum in these spaces is just the spectrum in L^∞ shifted by γ_c to the right.

The Proposition 26.1 is a partial answer to the selection problem: It excludes the slowest fronts because they are unstable in their moving frame. We also observe that this result is valid for fronts which are not positive while it does not discriminate among the fronts travelling at speed faster than 2. In order to complete the selection we need a more delicate argument, which is similar to the one used in Section 16 for convective instabilities.

We consider only the linearized problem around the zero solution and initial perturbations v with compact support. The evolution equation for these perturbations is then

$$\partial_t u(x,t) = \partial_x^2 u(x,t) + u(x,t) ,$$

with $u(x,0) = v(x)$. Calling w the Fourier transform of v, we have

$$u(x,t) = \int d\xi \, w(\xi) e^{i\xi x + \sigma(\xi)t} , \tag{26.1}$$

where $\sigma(\xi) = 1 - \xi^2$. Note that w is entire since v has compact support. We are interested on the apparent behavior of this solution for an observer moving with speed c, that is, we look at $u(ct + x_0, t)$. For simplicity, we set $x_0 = 0$. As $t \to \infty$ the behavior of $u(ct, t)$ is given by the exponential contribution at the stationary phase in the integral in (26.1). The critical point for $i\xi ct + \sigma(\xi)t$ is the solution ξ_0 of

$$ic + \sigma'(\xi) = 0 .$$

In our particular case, we find $\xi_0 = ic/2$. The growth rate of $u(ct, t)$ is then given by

$$\mathrm{Re}\,(ic\xi_0 + \sigma(\xi_0)) .$$

This growth rate is zero only when the speed c is equal to 2. Thus, in this linearized approximation, one can argue that an observer sees a stationary amplitude when travelling at speed $c = 2$. If he is slower, he sees a growing amplitude, and if he is faster, he outruns the solution. This second stability result completes the marginal stability criterion for the RA. It should be noted that this argument, while very elegant, does not show compellingly that an initial condition with compact support grows with speed $c = 2$. It only shows, that neglecting nonlinearities, the initial condition does not seem to go faster than speed $c = 2$.

27. Differential Operators. Nullspaces and Inverses

In this section, we shall study the nullspaces and inverses of several differential operators. In other words, we give bounds on the solutions of differential equations. In later developments, we need these bounds in the form of operator norms, because, as in the theory of bifurcations in Section 17.2, we want to iterate these operators. Then a formulation in terms of operator norms in suitable spaces seems most adequate. Throughout this section, we consider the inverse only on the half-space $x > 0$. This means that the exponential function, e^{-x} is *bounded*, in contrast to the problem on the full space $x \in [-\infty, \infty]$.

27.1. Nullspaces

We start with the simplest example,

$$(-i\partial_x - a)g(x) = 0 . \tag{27.1}$$

The solutions are

$$g(x) = \text{const. } e^{iax} .$$

The second example is

$$(-i\partial_x - a)^k g(x) = 0 , \tag{27.2}$$

for $k > 1$. Then, the solutions are linear combinations of

$$g(x) = x^j e^{iax} , \quad j = 0, \dots, k-1 , \tag{27.3}$$

as is easily verified. Below, we shall treat the case of a general differential polynomial and of matrices of differential polynomials.

27.2. Bounds on the Inverse

In several situations, we will need to solve differential equations of the form

$$(-i\partial_x - a)g(x) = f(x) , \tag{27.4}$$

where a is some constant, f is given and g is the unknown function. We will also need to discuss the more general equation

$$(-i\partial_x - v(x))g(x) = f(x) . \tag{27.5}$$

It is useful to define a linear operator L_v by

$$(L_v g)(x) \equiv (-i\partial_x - v(x))g(x) ,$$

so that the problem (27.5) amounts to inverting L_v. We are interested in solutions to this problem on *half-spaces*, and we take for the time being the half-space $x \ge 0$. The solutions of equation (27.5) are of one of the two forms

$$\begin{aligned} g(x) &= \frac{-i}{u(x)}\bigl(\text{const.} + \int_x^\infty dy\, f(y)u(y)\bigr) , \\ g(x) &= \frac{+i}{u(x)}\bigl(\text{const.} + \int_0^x dy\, f(y)u(y)\bigr) , \end{aligned} \tag{27.6}$$

where u is given by

$$u(x) = \exp\bigl(-i\int_0^x dy\, v(y)\bigr) . \tag{27.7}$$

To verify this assertion, it suffices to consider first the equation

$$\bigl(-i\partial_x - i(\log u)'(x)\bigr)\, g(x) = f(x) ,$$

which is obviously solved by (27.6). Then one observes that (27.7) solves

$$(\log u)'(x) = -iv(x) .$$

Note that the integration constant in (27.7) drops out in (27.6).

We next analyze the existence question for this problem. To do this, we define spaces of exponentially decaying functions on $[0, \infty)$.

Definition. For every $\gamma \in \mathbf{R}$, we define the space $\mathbf{H}_\gamma$ by

$$\mathbf{H}_\gamma = \left\{ f : \mathbf{R}^+ \to \mathbf{C} \ : \ \sup_{x\geq 0} |f(x)e^{\gamma x}| < \infty \right\} .$$

We also define, for $f \in \mathbf{H}_\gamma$, the norm

$$\|f\|_{\mathbf{H}_\gamma} = \sup_{x\geq 0} |f(x)e^{\gamma x}| .$$

We shall first concentrate on the case $v(x) = a$, with some (real or complex) constant a. Then we have $u(x) = e^{-iax}$. We then discuss the situation where $v(x)$ deviates from a constant by an integrable function.

Lemma 27.1. *Assume $\gamma \in \mathbf{R}$ is given. If $\gamma - \operatorname{Im} a > 0$, then the operator $L_a = -i\partial_x - a$ has an inverse which maps $\mathbf{H}_\gamma$ to itself. This inverse, g, is given by*

$$g(x) = -ie^{iax} \int_x^\infty dy\, e^{-iay} f(y) ,$$

and

$$\|g\|_{\mathbf{H}_\gamma} \leq \frac{\|f\|_{\mathbf{H}_\gamma}}{\gamma - \operatorname{Im} a} .$$

If $\gamma - \operatorname{Im} a < 0$, an inverse is given by

$$g(x) = ie^{iax} \int_0^x dy\, e^{-iay} f(y) ,$$

and

$$\|g\|_{\mathbf{H}_\gamma} \leq \frac{\|f\|_{\mathbf{H}_\gamma}}{|\gamma - \operatorname{Im} a|} .$$

Proof. According to (27.7), we have, for $x \geq 0$, and under the condition $\gamma - \operatorname{Im} a > 0$, the inequality

$$\begin{aligned} e^{\gamma x}|g(x)| &\leq e^{\gamma x} e^{-x \operatorname{Im} a} \int_x^\infty dy\, e^{y \operatorname{Im} a} e^{-\gamma y} \|f\|_{\mathbf{H}_\gamma} \\ &= \frac{1}{\gamma - \operatorname{Im} a} \|f\|_{\mathbf{H}_\gamma} . \end{aligned}$$

When $\gamma - \operatorname{Im} a < 0$, we have the bound

$$\begin{aligned} e^{\gamma x}|g(x)| &\leq e^{\gamma x} e^{-x \operatorname{Im} a} \int_0^x dy\, e^{y \operatorname{Im} a} e^{-\gamma y} \|f\|_{\mathbf{H}_\gamma} \\ &\leq e^{(\gamma - \operatorname{Im} a)x} \left(e^{-(\gamma - \operatorname{Im} a)x} - 1\right) \frac{1}{\operatorname{Im} a - \gamma} \|f\|_{\mathbf{H}_\gamma} \\ &= \frac{1}{|\gamma - \operatorname{Im} a|} \|f\|_{\mathbf{H}_\gamma} . \end{aligned}$$

We next note the important fact that L_a^{-1} is regularizing, since it is the inverse of a differential operator.

Lemma 27.2. *If the first k derivatives of f are in $\mathbf{H}_\gamma$, and if $g = L_a^{-1} f$, then the first $k+1$ derivatives of g are in $\mathbf{H}_\gamma$, provided $|\gamma - \operatorname{Im} a| > 0$. One has the bound*

$$\|g'\|_{\mathbf{H}_\gamma} \leq \left(\frac{|a|}{|\gamma - \operatorname{Im} a|} + 1\right) \|f\|_{\mathbf{H}_\gamma} . \tag{27.8}$$

Proof. It clearly suffices to deal with the case $k = 0$. The other cases follow by considering $\partial_x^k f$ instead of f. By construction, we have

$$-ig'(x) = f(x) + ag(x) ,$$

so that

$$\|g'\|_{\mathbf{H}_\gamma} \leq \|f\|_{\mathbf{H}_\gamma} + |a|\, \|g\|_{\mathbf{H}_\gamma} ,$$

from which the assertion follows.

We next deal with the case when a is not constant. The setting in which we want to work is described by the following assumption on v:

Assumption. We assume v is of the form

$$v(x) = a + w(x) ,$$

with $w \in \mathrm{L}^1 \cap \mathrm{L}^\infty$, that is, $\|w\|_1 \equiv \int_0^\infty dy\, |w(y)| < \infty$ and $|w(x)| \leq$ const. . In this case, we shall say that v is *nearly constant.*

Lemma 27.3. *Assume that v is nearly constant, with a satisfying $|\gamma - \operatorname{Im} a| > 0$. Then the inverse of the operator $L_v = -i\partial_x - v(x)$,* *defined by (27.6), (27.7), with the constant equal to zero, maps* $\mathbf{H}_\gamma$ *to itself, and* $g = L_v^{-1} f$ *satisfies the bounds*

$$\|g\|_{\mathbf{H}_\gamma} \leq \frac{e^{2\|w\|_1}}{|\gamma - \operatorname{Im} a|} \|f\|_{\mathbf{H}_\gamma} ,$$

$$\|g'\|_{\mathbf{H}_\gamma} \leq \left(\frac{(|a| + \|w\|_\infty) e^{2\|w\|_1}}{|\gamma - \operatorname{Im} a|} + 1 \right) \|f\|_{\mathbf{H}_\gamma} .$$

Remark. It is easy to give similar bounds when the integration constant in (27.6) is not zero.

Proof. We treat only the case when $\gamma - \operatorname{Im} a > 0$, the other case is handled equivalently. We begin by bounding the function u. Because v is nearly constant, we have

$$u(x) = \exp\left(-iax - i \int_0^x dy\, w(y)\right) .$$

This leads to the bound

$$e^{x \operatorname{Im} a - \|w\|_1} < |u(x)| < e^{x \operatorname{Im} a + \|w\|_1} .$$

Substituting these bounds, we get

$$\|g\|_{\mathbf{H}_\gamma} \leq \sup_{x \geq 0} \left(e^{(\gamma - \operatorname{Im} a)x} e^{\|w\|_1} \int_x^\infty dy\, e^{-(\gamma - \operatorname{Im} a)y} e^{\|w\|_1} \|f\|_{\mathbf{H}_\gamma} \right)$$

$$\leq \frac{e^{2\|w\|_1}}{\gamma - \operatorname{Im} a} \|f\|_{\mathbf{H}_\gamma} .$$

The bound on $\|g'\|_{\mathbf{H}_\gamma}$ is now straightforward, and follows as in the proof of Lemma 27.2.

27.3. Nullspaces and Inverses for Higher Order Operators

Nullspaces. The results of Section 27.2 apply to higher order operators as well. It suffices to write the higher order differential operator

$$P(-i\partial_x) = \sum_{j=0}^{n} a_j(-i\partial_x)^j$$

as a product:

$$P(p) = P_0 \prod_{j=1}^{n}(p - p_j) ,$$

where the p_j are the zeroes of the polynomial $P(p) = \sum_{j=0}^{n} a_j p^j$ and $P_0 \neq 0$. One then successively applies the operators $(-i\partial_x - p_j)$, as in (27.4). The nullspaces are obtained immediately from (27.3).

Definition of the Inverse for Simple Roots. We now turn to another representation which is very useful in the case of almost degenerate roots and degenerate roots. We begin with the case of simple roots. We fix a constant $\gamma \in \mathbf{R}$, and we assume that

$$\begin{aligned} \gamma - \operatorname{Im} p_1 &> 0 , \ldots , \gamma - \operatorname{Im} p_k > 0 , \\ \gamma - \operatorname{Im} p_{k+1} &< 0 , \ldots , \gamma - \operatorname{Im} p_n < 0 . \end{aligned} \tag{27.9}$$

By assumption, the roots p_j are all distinct. We denote by $\mathbf{L}$ the linear operator $P(-i\partial_x)$, that is,

$$\mathbf{L} = P_0 \prod_{j=1}^{n}(-i\partial_x - p_j) .$$

An inverse of $\mathbf{L}$ can then be defined by

$$\begin{aligned} (\mathbf{L}^{-1}f)(x) = & -i\sum_{j=1}^{k} \int_x^{\infty} dy \, \frac{e^{ip_j(x-y)}}{P'(p_j)} f(y) \\ & +i \sum_{j=k+1}^{n} \int_0^{x} dy \, \frac{e^{ip_j(x-y)}}{P'(p_j)} f(y) . \end{aligned} \tag{27.10}$$

We begin by checking that (27.10) defines an inverse, at least on a formal level. We then address the question of adequate function spaces. To see that (27.10) really defines an inverse, we apply $\mathbf{L}$ to the r.h.s. of (27.10). Note that for any constant a, we have,

$$(-i\partial_x - a)i\int_0^x dy\,\frac{e^{ip_j(x-y)}}{P'(p_j)}f(y)$$
$$= \frac{f(x)}{P'(p_j)} + (p_j - a)i\int_0^x dy\,\frac{e^{ip_j(x-y)}}{P'(p_j)}f(y)\,.$$

We see that for the particular choice of $a = p_j$, the second term on the r.h.s. drops. Therefore, we find, for every j,

$$\begin{aligned} &P_0\prod_{\ell=1}^{n}(-i\partial_x - p_\ell)i\int_0^x dy\,\frac{e^{ip_j(x-y)}}{P'(p_j)}f(y)\\ &\quad= P_0\prod_{\ell:\ell\neq j}\frac{(-i\partial_x - p_\ell)f(x)}{P'(p_j)}. \end{aligned} \tag{27.11}$$

Similarly, we find

$$\begin{aligned} &-P_0\prod_{\ell=1}^{n}(-i\partial_x - p_\ell)\cdot i\int_x^\infty dy\,\frac{e^{ip_j(x-y)}}{P'(p_j)}f(y)\\ &\quad= P_0\prod_{\ell:\ell\neq j}\frac{(-i\partial_x - p_\ell)f(x)}{P'(p_j)}. \end{aligned} \tag{27.12}$$

Taking now the Fourier transform of (27.11), we see that the contribution from the term j is

$$P_0\frac{\prod_{\ell\neq j}(p - p_\ell)\tilde{f}(p)}{P'(p_j)}\,.$$

We now claim that

$$\sum_{j=1}^{n} P_0\frac{\prod_{\ell\neq j}(p - p_\ell)}{P'(p_j)} = 1\,. \tag{27.13}$$

This clearly will suffice to show that (27.10) really defines an inverse. To prove (27.13), we use contour integration, namely, the identity, valid for sufficiently large r:

$$0 = \frac{1}{2\pi i}\oint_{|z|=r}\frac{dz}{(p - z)P(z)} = \sum_{j=1}^{n}\frac{1}{(p - p_j)P'(p_j)} - \frac{1}{P(p)}\,. \tag{27.14}$$

Rearranging terms, we see that (27.13) holds.

We next set out to prove bounds on the inverse as defined above. This is done as is the proof of Lemma 27.1. We again work in the spaces $\mathbf{H}_\gamma$. When $\gamma - \operatorname{Im} p_j > 0$, the same methods as before lead us to bound the integral

$$\int_x^\infty dy\, e^{\gamma(x-y) - \operatorname{Im} p_j (x-y)} .$$

When $\gamma - \operatorname{Im} p_j < 0$ we are led to bound

$$\int_0^x dy\, e^{\gamma(x-y) - \operatorname{Im} p_j (x-y)} .$$

These integrals exist, and are uniformly bounded in $x > 0$. We can summarize the result as follows:

Lemma 27.4. *If γ satisfies the inequalities (27.9) for $j = 1, \dots, n$, then the operator $\mathbf{L}^{-1}$ is a bounded map from $\mathbf{H}_\gamma$ to itself, whose norm is bounded by*

$$\|\mathbf{L}^{-1}\|_{\mathbf{H}_\gamma \to \mathbf{H}_\gamma} \le \sum_{j=1}^n \frac{1}{|(\gamma - \operatorname{Im} p_j) P'(p_j)|} .$$

Lemma 27.5. *Under the assumptions of Lemma 27.4, one has the bounds*

$$\|\partial_x^m (\mathbf{L}^{-1} f)\|_{\mathbf{H}_\gamma} \le \sum_{j=1}^n \frac{(1 + |p_j|)^m}{|(\gamma - \operatorname{Im} p_j) P'(p_j)|} \|f\|_{\mathbf{H}_\gamma} ,$$

for $m = 1, \dots, n$, where n is the degree of the differential operator. Furthermore, the number of free parameters for the inverse is equal to the number of indices j for which $\gamma - \operatorname{Im} p_j \le 0$.

Multiple Roots. The identities we have derived above generalize, without too much pain, to the case of polynomials with multiple roots. Let P be a polynomial of degree n, and let $q_1, \dots, q_j$ be the distinct roots of P, and $m_1, \dots, m_j$ their multiplicities. We have already discussed the nullspaces for the operator $P(-i\partial_x)$, they are spanned by the functions

$$x^m e^{i q_k x} , \quad \text{for } k = 1, \dots, j , \quad m = 0, \dots, m_k - 1 .$$

To compute the inverse, we use the Cauchy formula in the following form:

$$\frac{1}{2\pi i}\oint dz\,\frac{f(z)}{(z-q)^{m+1}} = \frac{1}{m!}\,\partial_z^m f(z)\big|_{z=q}\,,$$

or equivalently with $g(z) = f(z)/(z-q)^{m+1}$,

$$\frac{1}{2\pi i}\oint dz\, g(z) = \frac{1}{m!}\,\partial_q^m\,\left((z-q)^{m+1}g(z)\right)\big|_{z=q}\,.$$

We apply this now to the polynomial P above and obtain, for sufficiently large r,

$$\begin{aligned} 0 &= \frac{1}{2\pi i}\oint_{|z|=r}\frac{dz}{(p-z)P(z)} \\ &= -\frac{1}{P(p)} + \sum_{k=1}^{j}\frac{1}{(m_k-1)!}\,\partial_z^{m_k-1}\left(\frac{(z-q_k)^{m_k}}{(p-z)P(z)}\right)\bigg|_{z=q_k}\,. \end{aligned}$$

If $m_k = 1$, this leads to terms of the form $1/\big((p-q_k)P'(q_k)\big)$, as before. Going to Fourier transforms, we see that an inverse of the operator $\mathbf{L} = P(-i\partial_x)$ is given by

$$\begin{aligned} \mathbf{L}^{-1}f(x) &= \sum_{k=1}^{j}\frac{\pm i}{(m_k-1)!} \\ &\quad\times\int dy\, f(y)\partial_z^{m_k-1}\left(e^{iz(x-y)}\frac{(z-q_k)^{m_k}}{P(z)}\right)\bigg|_{z=q_k}\,. \end{aligned}$$

We want to work again in the spaces $\mathbf{H}_\gamma$. The integration limits are the same as in (27.10): If $\gamma - \operatorname{Im} q_k > 0$, we integrate from x to ∞ and choose the $-$ sign, otherwise, we integrate from 0 to x and choose the $+$ sign.

27.4. Coupled Systems

The theory of the preceding sections carries over to coupled systems of operators. A general formulation is possible in terms of eigenvectors, Jordan normal forms, and the like. The notation is, however, cumbersome in full generality. Our main interest in coupled systems is connected with reality requirements for solutions. We therefore only consider the case of two-component problems.

We consider two polynomials $P(x)$ and $Q(x)$. We are looking for solutions of the differential equation

$$P(-i\partial_x)f(x) + Q(-i\partial_x)\bar{f}(x) = 0 . \qquad (27.15)$$

Equations such as (27.15) occur for example in the bifurcation analysis of Section 17.3 in the following form: The operators P and Q are

$$P_n(z) = C_1 - (1-(n\omega+z)^2)^2 ,$$
$$Q(z) = C_2 ,$$

where C_1 and C_2 are real constants, and ω is real. We want to solve the problem

$$P_1(-i\partial_x)f_1(x) + Q(-i\partial_x)f_{-1}(x) = 0 ,$$
$$P_{-1}(-i\partial_x)f_{-1}(x) + Q(-i\partial_x)f_1(x) = 0 ,$$

with the additional condition

$$f_{-1}(x) = \bar{f}_1(x) .$$

Note now that $P_{-1}(-\bar{z}) = \big(P_1(z)\big)^-$, so that the two equations coincide and lead to a problem of the form of (27.15). Recall that $(\cdot)^-$ denotes the complex conjugate.

Lemma 27.6. *Let p be a simple zero of the equation*

$$P(p)\big(P(-\bar{p})\big)^- Q(p)\big(Q(-\bar{p})\big)^- = 0 . \qquad (27.16)$$

Then the functions

$$f(x) = C\big(P(-\bar{p})\big)^- e^{ipx} - \bar{C}Q(-\bar{p})e^{-i\bar{p}x}$$

are solutions of (27.15) for all $C \in \mathbf{C}$. If $\bar{P}(-\bar{p})$ and $Q(-\bar{p})$ are both zero, then choose f in the form

$$f(x) = CQ(p)e^{ipx} - \bar{C}\big(P(p)\big)^{-}e^{-i\bar{p}x} .$$

There are no other solutions of exponential form. In the case of multiple zeros p, the solutions are suitable combinations of $x^j \exp(ipx)$ and $x^{j'} \exp(-i\bar{p}x)$.

Proof. Setting $C = 1$ for simplicity, we see that

$$\begin{aligned} &P(-i\partial_x)f(x) + Q(-i\partial_x)\bar{f}(x) \\ &\quad = P(p)\big(P(-\bar{p})\big)^{-} e^{ipx} - P(-\bar{p})Q(-\bar{p})e^{-i\bar{p}x} \\ &\quad + Q(-\bar{p})P(-\bar{p})e^{-i\bar{p}x} - Q(p)\big(Q(-\bar{p})\big)^{-} e^{ipx} , \end{aligned}$$

and this quantity is zero by (27.16). If we write an ansatz

$$f(x) = \alpha e^{ipx} + \beta e^{-i\bar{p}x} ,$$

then we see that the condition (27.16) says exactly that the determinant of the matrix equation corresponding to (27.15) is zero. Hence, no other solutions exist. The last assertion follows by inspection.

After the search for the null spaces, we now want to derive an identity for the inverse on half-spaces. It is a generalization of (27.10). We discuss only a very special case, relevant to the developments in this book. Let g be a given complex function and consider the problem

$$\begin{aligned} P_1(-i\partial_x)f_1(x) + Q_1(-i\partial_x)f_{-1}(x) &= g(x) , \\ P_{-1}(-i\partial_x)f_{-1}(x) + Q_{-1}(-i\partial_x)f_1(x) &= h(x) . \end{aligned} \tag{27.17}$$

We want to construct solutions of (27.17). This approach is to be contrasted with earlier developments, where we imposed immediately the reality conditions; see Section 17.4.

We have now a matrix valued problem with matrix

$$\begin{pmatrix} P_1(p) & Q_1(p) \\ Q_{-1}(p) & P_{-1}(p) \end{pmatrix} ,$$

whose inverse is

$$\frac{1}{D(p)} \begin{pmatrix} P_{-1}(p) & -Q_1(p) \\ -Q_{-1}(p) & P_1(p) \end{pmatrix} ,$$

and

$$D(p) = P_1(p)P_{-1}(p) - Q_1(p)Q_{-1}(p) .$$

Let us assume that D has only simple zeroes at p_j, $j = 1, \ldots, n$. Then we can rewrite the Fourier transform of the solution of (27.17) as

$$\sum_{j=1}^{n} \frac{1}{(p-p_j)D'(p_j)} \begin{pmatrix} P_{-1}(p_j) & -Q_1(p_j) \\ -Q_{-1}(p_j) & P_1(p_j) \end{pmatrix} \begin{pmatrix} \tilde{g}(p) \\ \tilde{h}(p) \end{pmatrix} .$$

In the original variables, this leads to

$$\sum_{j=1}^{n} \pm i \int dy\, e^{ip_j(x-y)} \frac{1}{D'(p_j)} \begin{pmatrix} P_{-1}(p_j) & -Q_1(p_j) \\ -Q_{-1}(p_j) & P_1(p_j) \end{pmatrix} \begin{pmatrix} g(y) \\ h(y) \end{pmatrix} . \tag{27.18}$$

Again, the integration limits depend on the sign of $\gamma - \operatorname{Im} p_j$: If this quantity is positive, we integrate from 0 to x and choose the $+$ sign. If it is negative, we integrate from x to ∞ and choose the $-$ sign.

Assume in addition that

$$P_{-1}(\bar{z}) = \left(P_1(-z)\right)^{-}$$

and

$$Q_{-1}(\bar{z}) = \left(Q_1(-z)\right)^{-} .$$

Then one can show that $h(x) = \bar{g}(x)$ implies the existence of a solution for which $\bar{f}_1(x) = f_{-1}(x)$.

27.5. The Number of Solutions

Counting the correct number of solutions in bifurcation problems is a task of utmost theoretical and practical importance. We have seen this in Section 21 where we constructed solutions to bifurcation problems by "patching together" solutions for $x < 0$ with solutions for $x > 0$. We will also use this in the construction of fronts. Thus, we need to know the number of free parameters which remain on each side of $x = 0$. This number of parameters can be efficiently computed from the representation (27.6), together with the bounds of Lemma 27.1.

We only discuss the operator $(-i\partial_x - a)$ as defined in (27.4), leaving to the imagination of the reader any variants of this case. In this case, an eigenfunction v with eigenvalue 0 is given by

$$v(x) = e^{iax} . \tag{27.19}$$

Assume now that γ is given. Then we see that $v \in \mathbf{H}_\gamma$ if and only if $\gamma - \operatorname{Im} a \leq 0$. Thus, in the case $\gamma - \operatorname{Im} a \leq 0$, there is a one parameter family of solutions $g \in \mathbf{H}_\gamma$ of $(-i\partial_x - a)g(x) = f(x)$ given by

$$g(x) = \xi e^{iax} + i \int_0^x dy\, e^{ia(x-y)} f(y) \ , \quad \xi \in \mathbf{C} \ .$$

On the other hand, if $\gamma - \operatorname{Im} a \geq 0$, there is only the solution

$$g(x) = -i \int_x^\infty dy\, e^{ia(x-y)} f(y) \ .$$

28. Fronts for the Swift-Hohenberg Equation

We consider the SH equation and we formulate first an adequate definition of fronts for this case. The idea is basically that a front solution describes a solution which, starting with amplitude 0, grows in time and then leaves a stationary solution in the *laboratory frame*. This is analogous to the definition we gave for the CA equation in (24.4). We have seen that, for small values of the bifurcation parameter, the stationary solutions of the SH equation are essentially of the form $2\varepsilon \cos(\omega x)$. Therefore, they are *not* translation invariant in the laboratory frame. It follows that a front cannot be expected to be a solution of the form $U(x,t) = V(x - ct)$. In fact, we have seen that fronts for the RA equation are of this form but it should now be clear that this is an accidental degeneracy due to the translation invariance of the stationary solution.

We therefore seek a solution which really depends on space and time, but we are free to choose the precise functional dependence for this solution, provided the variables are independent. We shall select the following two coordinates:

- The laboratory coordinate x_L.
- The front coordinate x_F, which moves with constant speed c relative to the laboratory coordinate.

We expect the following picture to emerge: Consider a front solution $W(x_L, x_F)$, and assume that the front is moving from left to right. If we fix x_L, and consider the limit, as x_F goes to ∞, we are in fact looking at the behavior at a fixed laboratory point of the solution as the front

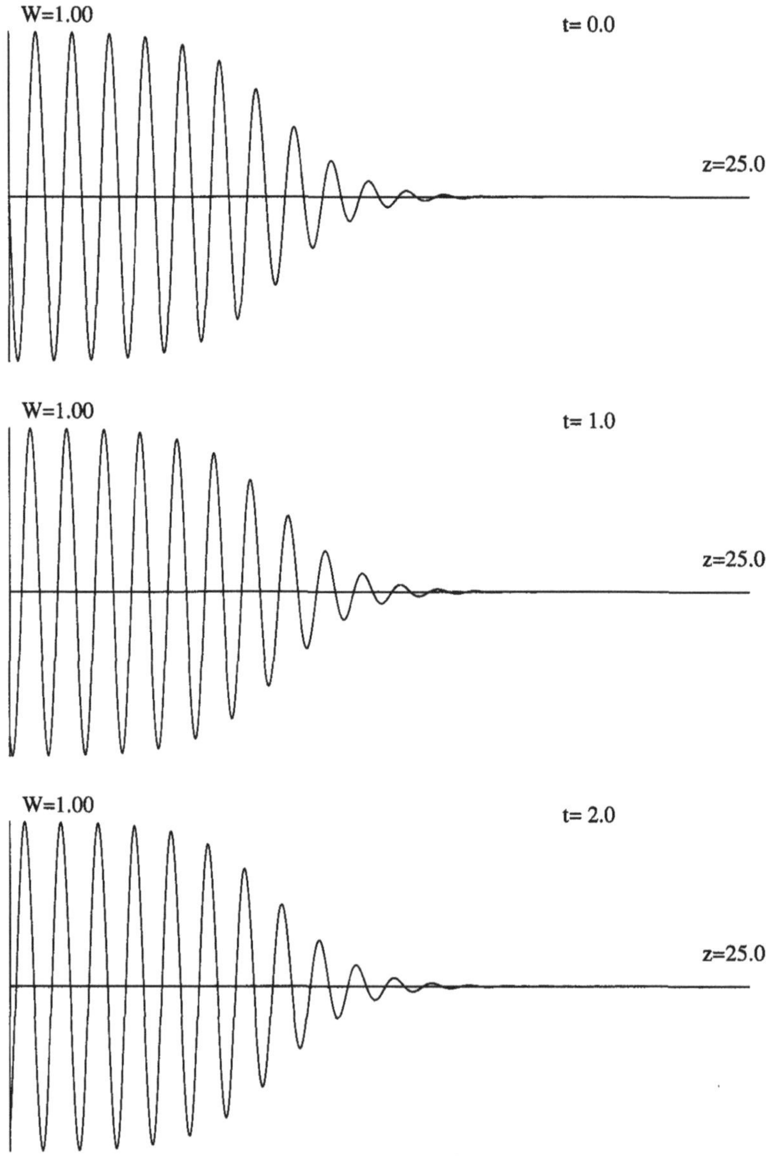

Fig. 29: The front for the SH equation in a frame moving with the envelope. Note that in this frame, the oscillations seem to move backward inside the envelope. The time step is one third of a period.

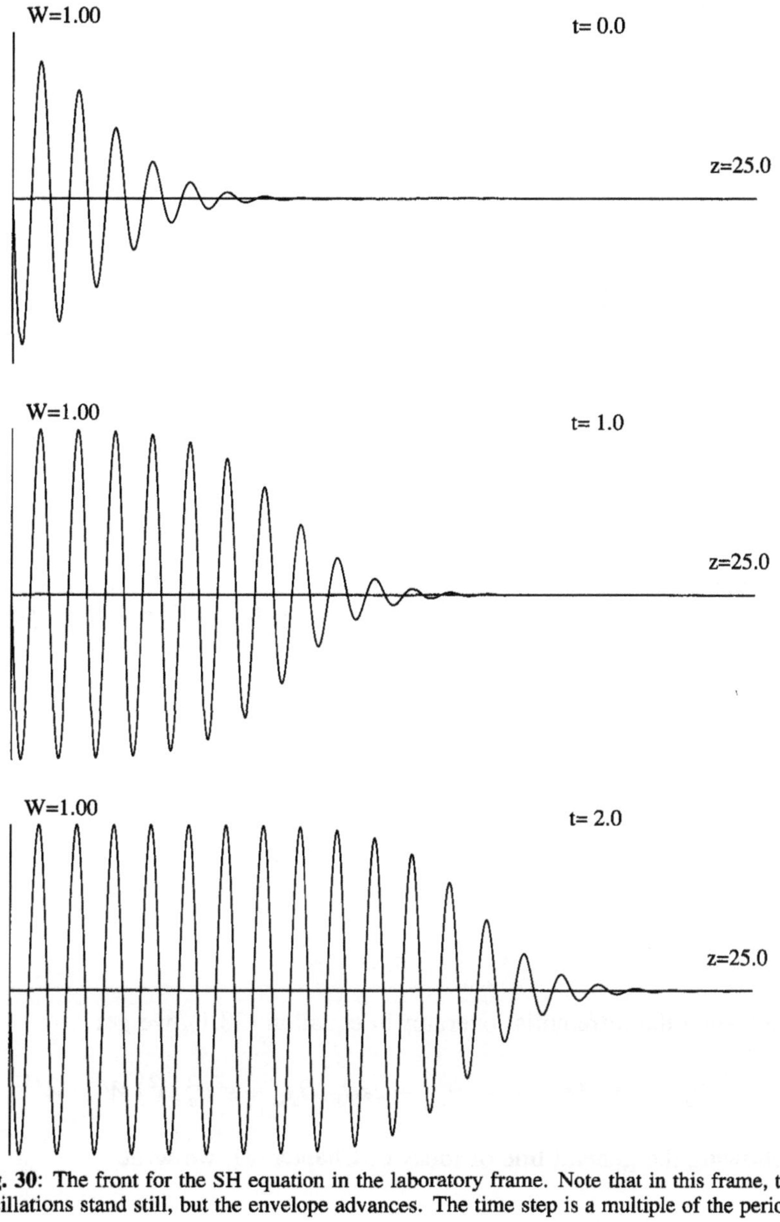

Fig. 30: The front for the SH equation in the laboratory frame. Note that in this frame, the oscillations stand still, but the envelope advances. The time step is a multiple of the period.

passes by. Therefore, we expect $W(x, x_F)$ to be close to a translate of a stationary solution, when viewed as a function of x. More precisely, there should exist a $S(x)$ which is a stationary solution of the SH equation such that

$$W(x, x_F) - S(x) \to 0 \quad \text{as} \quad x_F \to \infty .$$

To understand the dependence on x_F, we assume that $W(x_L, x_F)$ is of the approximate form

$$W(x_L, x_F) \approx S(x_L) \cdot A(x_F) .$$

(We shall see that this approximation is qualitatively correct.) The function A is the amplitude (or envelope) which has a constant shape in the moving frame, and this function has the limits $A(-\infty) = 1$, $A(+\infty) = 0$. It therefore looks like a solution of the real amplitude equation, and we shall see that in perturbation theory the lowest order term indeed satisfies a second order semilinear equation, similar to the real amplitude equation. The picture which emerges is thus: A fixed envelope moves from left to right, and inside it, the solution $S(x_L)$ grows in the laboratory frame. This is illustrated in Fig. 29 and in Fig. 30.

We now start a multiscale analysis, leading to perturbation theory, and, based on it, to a proof of existence of a solution. We will, however, not go through all the details of this proof in this book.

We consider solutions of the form $W(x, \varepsilon x - \varepsilon^2 ct)$, and substitute into the equation

$$\partial_t W = \left(\alpha - (1 + \partial_x^2)^2\right) W - W^3 . \tag{28.1}$$

We fix a frequency ω and a bifurcation parameter ε, and we study the equation (28.1) in perturbation theory. We have the relation

$$\alpha = 3\varepsilon^2 + (1 - \omega^2)^2 .$$

Expanding the differential operator in equation (28.1), we get

$$-\varepsilon^2 c \partial_{x_F} W = \left(\alpha - (1 + \partial_{x_L}^2 + 2\varepsilon \partial_{x_L} \partial_{x_F} + \varepsilon^2 \partial_{x_F}^2)^2\right) W - W^3 .$$

Following the general line of ideas of Chapter VI, we write

$$W(x_L, x_F) = \varepsilon \sum_{n \in \mathbf{Z}} e^{i\omega n x_L} W_n(x_F) ,$$

where $W_{-n} = \bar{W}_n$. The case of constant W_n coincides exactly with our bifurcation analysis of Section 17.2. The function $W(x, \varepsilon x - \varepsilon^2 ct)$ is slowly varying in x and t in its second argument. (This is why we speak of a multiscale analysis.)

The equations for the functions $W_n(x)$ are now

$$\begin{aligned}\big(\alpha - (1 - n^2\omega^2 + 2i\varepsilon n\omega\partial_x + \varepsilon^2\partial_x^2)^2\big)W_n + \varepsilon^2 c\partial_x W_n(x) \\ = \varepsilon^2 \sum_{p+q+r=n} W_p(x)W_q(x)W_r(x) \,.\end{aligned}$$

To simplify the notation somewhat during the exploratory phase, we pursue the case $\omega = 1$, and set $\alpha = 3\varepsilon^2$. Then we get the equations

$$\begin{aligned}\big(3\varepsilon^2 - (1 - n^2 + 2i\varepsilon n\partial_x + \varepsilon^2\partial_x^2)^2\big)W_n + \varepsilon^2 c\partial_x W_n \\ = \varepsilon^2 \sum_{p+q+r=n} W_p W_q W_r \,.\end{aligned} \tag{28.2}$$

As in the case of bifurcation theory, the interesting equations occur for $n = \pm 1$. Upon expansion, they yield

$$\varepsilon^2(-\varepsilon^2\partial_x^4 \mp 4i\varepsilon\partial_x^3 + 4\partial_x^2 + c\partial_x + 3)W_{\pm 1} = \varepsilon^2 \sum_{p+q+r=\pm 1} W_p W_q W_r \,. \tag{28.3}$$

Guided by our bounds on the amplitudes η of Theorem 17.1 – which will certainly hold for very negative values of x, that is, behind the front – we pursue our analysis under the assumption that $W_{\pm 1} \approx \mathcal{O}(1)$ and that the other W_p are $\mathcal{O}(\varepsilon)$. However, we shall see – and this is typical for the multiscale analysis – that $\varepsilon\partial_x^3$ is to be considered of the same order as ∂_x^2. Therefore, we keep the third and fourth order derivative terms.

Then Eq. (28.3) reduces to

$$(-\varepsilon^2\partial_x^4 \mp 4i\varepsilon\partial_x^3 + 4\partial_x^2 + c\partial_x + 3)W_1 - 3W_1|W_1|^2 = 0 \,, \tag{28.4}$$

where we have used $W_1 = \bar{W}_{-1}$.

Note first that $W_1 = 1$ is a solution, as well as $W_1 = 0$. If we now set $\varepsilon = 0$, under the assumption that all derivatives of W_1 stay bounded, the equation reduces to the complex amplitude equation:

$$(4\partial_x^2 + c\partial_x + 3)W_1 - 3W_1|W_1|^2 = 0 \,.$$

Thus we expect:

The SH equation has moving front solutions of the approximate form

$$U(x,t) = S(x)A(\varepsilon x - \varepsilon^2 ct) , \qquad (28.5)$$

where ε *and* ω *satisfy the relation*

$$3\varepsilon^2 + (1-\omega^2)^2 = \alpha ,$$

and

$$S(x) \approx 2\varepsilon \cos(\omega x) .$$

Furthermore, $A(z)$ *is a positive solution of the equation*

$$4A'' + cA' + 3A - 3A^3 = 0 ,$$

satisfying $A(-\infty) = 1$, $A(\infty) = 0$.

A change of variables transforms the last equation to the form we used earlier

$$B'' + \hat{c}B' + B - B^3 = 0 . \qquad (28.6)$$

Choose $B(\beta z) = A(z)$ with $\beta = \sqrt{3}/2$, and $\hat{c} = \beta c/3$. We have seen in Section 24 that Eq. (28.6) has positive solutions for $\hat{c} \geq 2$, and thus *we expect positive amplitudes for* $c \geq 4\sqrt{3}$.

The above statement can be proved [CE1]. Even after the preparations of the preceding parts of this book, this proof is too technical to be presented here. On the other hand, the reader should have acquired the tools which make the outline of the proof given below extremely plausible.

We first state the result in precise form.

Theorem 28.1. *There is an* $\varepsilon_0 > 0$ *and a* $\delta > 0$ *such that for every* ω *satisfying* $|1-\omega^2| < \delta$ *and for every* ε *satisfying* $0 < \varepsilon < \varepsilon_0$ *the equation (28.1) has, for every* $c > 4\sqrt{3}$ *sufficiently close to this limit, a moving front solution of the approximate form (28.5).*

The proof is a combination of the techniques developed in this book. The strategy consists of separating the proof into two main arguments, one valid *ahead* of the front, and the other valid in the *bulk*. We shall show that the equation has a solution in each of these two regions. Ahead of the front we use stable manifold theory (cf. Section 9), and in the bulk we use perturbation theory (for a 4th order operator) around the periodic solution. The two solutions are then matched together by a method which generalizes transversality ideas.

Remark. One can also perform a linear stability analysis for the fronts obtained above. See [DL], and [CE2]. This analysis is quite similar in spirit to a combination of the methods used to show the existence of stationary solutions in half-spaces, and of the methods for studying the stability of periodic solutions. While we have given these methods in detail, we do not describe the stability analysis for the SH equation in this book.

28.1. Analysis of the Solution Ahead of the Front

In this section, we study *small* solutions of the SH equation. The argument ahead of the front is a generalization of that given for the real amplitude equation in Section 24. There, we treated ($p = 0, q = 0$) as a fixed point of the corresponding Hamiltonian system. The "trajectory" in phase space is then the stable manifold of this fixed point. In the case of the RA equation, the phase space is 2-dimensional. In the case of the SH equation, the phase space is infinite dimensional, but the basic idea remains the same. However, even for small amplitudes, the 4th order operator is not a small perturbation, and has an important influence on the spectrum; see below. The unknowns are the amplitudes W_n, $n \in \mathbf{Z}$, with the restrictions $W_{-n} = \bar{W}_n$. Since the equations (28.2) are 4th order differential equations, each W_n is determined by 4 complex numbers. If we fix an $X \in \mathbf{R}$, then the 4 numbers are $W_n(X)$ and the first 3 derivatives of W_n at this point. Thus, we are in a situation which is a generalization of that encountered in Section 9.

In order to apply the stable manifold theorem, we analyze the spectrum of the linearized problem. Considering the equation (28.2), we see that the linearization leads to *decoupled* equations. Thus, we only have to consider the spectrum of each of the differential operators

$$P_{n\omega}(-i\partial_x) = \big(3\varepsilon^2 - (1 - n^2\omega^2 + 2i\varepsilon n\omega\partial_x + \varepsilon^2\partial_x^2)^2\big) + \varepsilon^2 c\partial_x \ ,$$

for $n \in \mathbf{Z}$. (We are only interested in values of ω close to 1.)

Again, as in the existence proof for stationary fronts, or as in the proof of Theorem 20.2, the sectors $n = \pm 1$ are more complicated than the other ones. When $n \neq \pm 1$, and n is odd, the polynomial $P_{n\omega}(p)$ has 2 roots with positive imaginary part and two roots with negative imaginary part. One can show [CE1, Lemma 7.1] that the roots $p_{n,j}, j = 1, \ldots, 4$ satisfy the bounds (for $n\omega > 0$)

$$|p_{n,j}| = \mathcal{O}(\frac{n\omega}{\varepsilon}) ,$$

$$|\mathrm{Im}\, p_{n,j}| \ge \mathcal{O}(1) \cdot \begin{cases} A & \text{if } n\omega \le \varepsilon^{-2/3} \\ \min(A, B) & \text{if } \varepsilon^{-2/3} \le n\omega \le \varepsilon^{-4/3} \\ B & \text{if } n\omega \ge \varepsilon^{-4/3} \end{cases} ,$$

where $A = (n\omega/\varepsilon)^{1/2}$ and $B = (n\omega/\varepsilon^3)^{1/4}$.

These inequalities are obtained readily by an asymptotic analysis of the polynomial. This analysis also leads to the following bounds on the derivative of $P_{n\omega}$ at the roots :

$$|P'_{n\omega}(p_{n,j})| \ge \mathcal{O}(n\omega^{1/2}\varepsilon^{3/2}) .$$

The above estimates are by no means optimal, but they are sufficiently good for our purpose. To apply the stable manifold theorem, we need a function space. Le f be a function which is C^4. To f, we can associate the four coordinates

$$f(X) , -if'(X) , (-i)^2 f''(X) , (-i)^3 f'''(X) \tag{28.7}$$

(these are four complex numbers). In this basis, the eigenvectors of the operator $P_{n\omega}(-i\partial_x)$ are of the form

$$\left(1 , p_{n,j} , p_{n,j}^2 , p_{n,j}^3\right) \in \mathbf{C}^4 , \tag{28.8}$$

and hence the stable and unstable eigenspaces E_n^s, E_n^u of $P_{n\omega}(-i\partial_x)$ are spanned each by two vectors of the above form. Note that, as functions of x, the eigenfunctions are of the form $e^{ip_{n,j}x}$, so that the roots with positive imaginary part correspond to the exponentially decaying solutions at $+\infty$.

Suppose now W_n is given. At $x = X$, we associate a quadruple to W_n, as in (28.7). Since the vectors in (28.8) form a basis of $\mathbf{C}^4$, we can write, for $k = 0, \ldots, 3$,

$$(-i)^k(\partial_x^k W_n)(X) = \sum_{j=1}^{4} b_{n,j} p_{n,j}^k ,$$

that is, this determines the coefficients $b_{n,j}$, $j = 1, \ldots, 4$. This construction defines a map from the set of all W_n to the set of all $b_{n,j}$ and we equip the set $\mathbf{B}$ of all such sequences $\mathbf{b}$ of $b_{n,j}$, $j = 1, \ldots, 4$, $n \in \mathbf{Z}$ with the norm

$$\|\mathbf{b}\|_{\mathbf{B}} = \sum_{n \in \mathbf{Z}} \varepsilon^{-pn} n^3 \max_{j=1,\ldots,4} |b_{n,j}| \ .$$

This defines a topology similar to that found in the study of stationary solutions.

Note. We have not yet handled the case $n = \pm 1$, but we continue the general discussion before dealing with that case.

One now checks that the linear operators and the nonlinearity satisfy the bounds required by Theorem 9.2. Therefore, that theorem applies and we have the following nice result (neglecting momentarily the problems with $n = \pm 1$).

In the space $\mathbf{B}$, there is, in a neighborhood of the origin, a stable manifold W^s which is tangent to the stable eigenspace E^s spanned by those $\mathbf{b}$ whose components $b_{n,j}$ are 0 whenever $\operatorname{Im} p_{n,j} < 0$. Translating this fact back to the W_n, we see that there is a 2 times infinite dimensional stable subspace. There are, loosely speaking, two relations among the four numbers $(-i\partial_x)^k W_n(X)$, $k = 0, \ldots, 3$. If they are satisfied for all n, then the coupled non-linear system (28.2) with initial conditions $(-i\partial_x)^k W_n(X)$, $k = 0, \ldots, 3$, has a solution which tends exponentially to zero for all n.

In the above discussion, we have neglected two things :

- We have not discussed the relation $W_n = \bar{W}_{-n}$.
- We have not considered the case $n = \pm 1$.

The first problem is easily solved by considering real instead of complex variables, and we give no details here.

The second problem is more interesting, because of the change in spectrum. (Note that at 0 amplitude, the cubic term does *not* imply a coupling of the type of (19.13).)

The remainder of this subsection is devoted to the analysis of the second case, that is, the sector $n = \pm 1$. To see what is going on, it is again instructive to consider the case $\omega = 1$, and it suffices to look at $n = 1$. Then

$$\begin{aligned} P_1(p) &= 3\varepsilon^2 - (+2\varepsilon p + \varepsilon^2 p^2)^2 + i\varepsilon^2 cp \\ &= \varepsilon^2 (3 - (2p + \varepsilon p^2)^2 + icp) \ . \end{aligned}$$

The limiting value of c is $c = 4\sqrt{3}$, and we restrict our attention to this value. Neglecting terms of order $\mathcal{O}(\varepsilon^3)$, the polynomial reduces to

$$P_1(p) = \varepsilon^2(3 + 4\sqrt{3}ip - 4p^2) ,$$

from which we see that P_1 has a double root at

$$p = i\sqrt{3}/2 .$$

(It is not astonishing that $c = 4\sqrt{3}$ leads to a double root, since the slowest speed for the real amplitude equation is characterized by this property.)

When ε is close to zero, we see that there are two other roots at about

$$p = -\frac{2}{\varepsilon} \pm \sqrt{\frac{-ic}{2\varepsilon}} .$$

Thus, for $n = 1$ there are 3 roots with positive imaginary part and 1 with negative imaginary part. The stable manifold in this sector will therefore be 3-dimensional; if one relation is satisfied among the four numbers $(-i\partial_x)^k W_1(X)$, $k = 0, \ldots, 3$, the solution with these initial conditions will go to zero exponentially (provided one also satisfies the corresponding relations in the other sectors).

28.2. Analysis in the Bulk

Here we analyze the equation near the stationary solution. This is a much harder problem than the analysis ahead of the front, because the nonlinearity cannot be neglected. On the other hand, it is still true that the only interesting problems occur for $n = \pm 1$. The equations for $n = \pm 1$ are again coupled, and taking up the discussion with equation (28.3), we are confronted with solving

$$(-\varepsilon^2\partial_x^4 - 4i\varepsilon\partial_x^3 + 4\partial_x^2 + c\partial_x + 3)W_1 - 3W_1|W_1|^2 = h , \qquad (28.9)$$

and its complex conjugate, with h the inhomogeneity

$$h = \sum_{\substack{p+q+r=1 \\ |p|+|q|+|r|>3}} W_p W_q W_r .$$

As we have said before, we expect h to be at least of order ε. We want to view (28.9) as a fixed point problem, and we study therefore in detail the nonlinear operator on the left-hand side. When ε is zero, this operator reduces, up to a rescaling, to the real amplitude equation. Assume now that $u(x)$ is a real solution of the amplitude equation :

$$4u'' + cu' + 3u - 3u^3 = 0 , \tag{28.10}$$

with $u(-\infty) = 1$, $u(+\infty) = 0$, $u(x) > 0$ for all $x \in \mathbf{R}$. This solution is unique up to translation invariance, which we break by imposing, for example, $u(0) = 1/2$. We now write

$$W_1(x) = u(x) + a(x) + ib(x) ,$$

where a and b are real functions, which we expect to be small when ε and h are small. Thus, we study the linearized problem, derived from (28.9) by expanding W_1

$$\begin{aligned} L_1 a &\equiv 4a'' + ca' + 3a - 9au^2 = h_1 , \\ L_2 a &\equiv 4b'' + cb' + 3b - 3bu^2 = h_2 . \end{aligned} \tag{28.11}$$

Here, we used

$$W_1|W_1|^2 = (u + a + ib)((u + a)^2 + b^2) ,$$

and we define

$$\begin{aligned} h_1 &= \operatorname{Re} h + 9a^2 u + 3a^3 u + \varepsilon^2 (a + u)^{iv} + 4\varepsilon b''' , \\ h_2 &= \operatorname{Im} h + 3b(2ua + a^2 + b^2) + \varepsilon^2 b^{iv} - 4\varepsilon (a + u)''' . \end{aligned}$$

The problem (28.9) has been transformed (for $\varepsilon = 0$) into a linear equation in a nonconstant "potential" (namely u^2), with a r.h.s. which contains an inhomogeneity, h, and the unknown functions a, b, which appear at least with the power 2. Thus, the problem is now transformed into

$$L\begin{pmatrix} a \\ b \end{pmatrix} = \begin{pmatrix} h_1(a, b) \\ h_2(a, b) \end{pmatrix} ,$$

where L is the operator from the l.h.s. of (28.11). One can check by a direct calculation that

$$\begin{aligned} L_1 f &= 4(\partial_x + \frac{u''}{u'} + \frac{c}{4})(\partial_x - \frac{u''}{u'}) f , \\ L_2 f &= 4(\partial_x + \frac{u'}{u} + \frac{c}{4})(\partial_x - \frac{u'}{u}) f , \end{aligned} \tag{28.12}$$

because u solves (28.10). Since u is a solution of a second order problem, the decay properties of u are extremely well-known, and one can therefore apply the methods of Section 27.3 to invert the operators L_1, and L_2 by quadratures. This allows us to reformulate the problem in the form

$$\begin{pmatrix} a \\ b \end{pmatrix} = \begin{pmatrix} L_1^{-1} h_1(a,b) \\ L_2^{-1} h_2(a,b) \end{pmatrix} ,$$

and it is this problem we solve as a fixed point problem. All of the above manipulations have been done for $\omega = 1$. If ω is close to, but not equal to 1, then we get additional terms in h_1 and h_2, but the main ideas remain the same.

We now analyze in more detail the operators h_1 and h_2. In particular, they contain terms of the form $\varepsilon^2 a^{iv}$, $\varepsilon^2 b^{iv}$ and $\varepsilon a'''$, $\varepsilon b'''$. These terms have a small coefficient, but the highest derivatives. This is a phenomenon which is perhaps better known in the study of the Schrödinger equation. For example, the eigenvalue problem is (after a rescaling of the mass)

$$-h^2 \partial_x^2 \psi(x) + V(x)\psi(x) = E\psi(x) ,$$

where h is Planck's constant. Since this is a small number, we see the analogy with the front problems for small ε. In quantum mechanics, the WKB method is used to desingularize the problem in the limit $h \to 0$. It consists in writing

$$\psi(x) = e^{\frac{i}{h} S(x)} \tag{28.13}$$

and solving for S instead of ψ. Note, however, the fundamental difference between the equations

$$\varepsilon^2 \xi^2 - 1 = 0 \tag{28.14}$$

and

$$\varepsilon^2 \xi^2 + \xi - 1 = 0 . \tag{28.15}$$

The first has two solutions which diverge as $\varepsilon \to 0$, while the second has one solution which tends to 1, and one solution which tends to ∞. (The difference between the two cases can be understood in terms of the Newton polygon.) In the first case, only WKB-like solutions of a form analogous to (28.13) appear, while in the second case one regular and one singular solution occurs. In the front problem, we will use the regular solution. The book by Maslov [M1] contains a multitude of examples where this choice must be made.

To continue with this study, we describe the result of acting with L_1^{-1} on, say, $\varepsilon^2 \partial_x^4$. A powerful methodology, the theory of pseudo-differential operators has been developed [H] to handle such problems. This method consists in an extension of Fourier transforms to functions (kernels) which depend on p and x. We do not need this machinery in its full power here, but it tells us at once that $L_1^{-1}\varepsilon^2\partial_x^4$ must be something like a second order differential operator (plus smoother terms), since L_1^{-1} is the *inverse* of a second order operator. A straightforward but tedious calculation involving commutators of differential operators show that

$$\begin{aligned} 4L_j^{-1}\partial_x^4 f &= f'' - \frac{c}{4}f' + R_{j4}(f)\ , \ j = 1,2\ , \\ 4L_1^{-1}\partial_x^3 f &= f' - \frac{c}{4}f + (\partial_x - u''/u')^{-1}R_{13}(f)\ , \\ 4L_2^{-1}\partial_x^3 f &= f' - \frac{c}{4}f + (\partial_x - u'/u)^{-1}R_{23}(f)\ . \end{aligned} \tag{28.16}$$

A space on which one can conveniently do estimates for such operators is H with a norm

$$\|f\|_H = \sup_{x\in(-\infty,X_\varepsilon]} |f(x)|e^{-\gamma x}\ ,$$

where $X_\varepsilon = \log(\frac{1}{\varepsilon})$ and γ satisfies $0 < \gamma < \beta$ with β the exponential rate at which $u(x)$ tends to 1 as $x \to -\infty$. The linear operators R_{ij} are then bounded in norm by $\mathcal{O}((\log|\varepsilon|)^5)$. A calculation like the one in Section 24 shows that

$$\beta = \frac{-c + \sqrt{c^2 + 96}}{8}\ .$$

To see this, write $u = 1 - v$ and substitute, and this leads to

$$-4v'' - cv' + bv + \mathcal{O}(v^2) = 0\ .$$

Setting $v = \exp(\beta x)$ the solution follows.

Before going on, we explain the choices of γ and X_ε which will be crucial for the matching to work. Recall that we have fixed the coordinate origin by requiring $u(0) = 1/2$. The question is now: How far to the right of $x = 0$ is it possible to control the analysis of the solution pretending it is close to 1? And what is the decay rate which we can expect to prove for the solution, given that the linearized problem has rate β? The decay rate for the nonlinear problem can be chosen as close to β as one wishes, if ε is sufficiently small. We find that X_ε is a judicious choice, because at this value of x the amplitude is already very small when ε is small. *This will be good enough to make contact with the region where the analysis ahead of the front can be shown to be valid.*

Going further to the right, for example to $X_\varepsilon = \varepsilon^{-1/2}$, would lead to bounds on the bulk solution which diverge too fast (like a power of ε^{-1} instead of $\log \varepsilon^{-1}$) and they could not be compensated by the factors ε^2 and ε in front of the differential operators.

We go back to the analysis of $L_j^{-1} h_j$, for $j = 1, 2$. Having found the correct function space, one can rewrite the problem as

$$\begin{aligned} a - L_1^{-1}(\varepsilon^2 \partial_x^4 a - 4\varepsilon \partial_x^3 b) &= h_3(a, b) \ , \\ b - L_2^{-1}(\varepsilon^2 \partial_x^4 b + 4\varepsilon \partial_x^3 a) &= h_4(a, b) \ , \end{aligned} \tag{28.17}$$

where the terms in h_3 are those from h_1 not containing a^{iv}, a''', b^{iv}, b''', and similarly for h_4 and h_2. In view of what we have said in (28.16) we see that the operator on the left-hand side of (28.17) is approximately equal to the differential matrix ($p = -i\partial_x$)

$$M = \begin{pmatrix} 1 + \frac{\varepsilon^2 p^2}{4} + \frac{i\varepsilon^2 pc}{4} & i\varepsilon p - \frac{\varepsilon c}{4} \\ -i\varepsilon p + \frac{\varepsilon c}{4} & 1 + \frac{\varepsilon^2 p^2}{4} + \frac{i\varepsilon^2 pc}{4} \end{pmatrix} .$$

The characteristic polynomial of M has (in p) the 4 roots

$$p = \begin{cases} 1 \pm \varepsilon^{1/2}(\frac{ic}{8})^{1/2} + \mathcal{O}(\varepsilon) \\ -1 \pm \varepsilon^{1/2} i(\frac{c}{8})^{1/2} + \mathcal{O}(\varepsilon) \end{cases} . \tag{28.18}$$

Thus, the inverse of M can be defined by the methods of Section 27 and one can check that

$$M^{-1} : H \oplus H \to H \oplus H$$

has norm $\mathcal{O}(\varepsilon^{-1})$, and $M^{-1}(\partial_x - \frac{u'}{u})^{-1}$ has norm $\mathcal{O}(\varepsilon^{-1/2})$. Since all terms on the right of (28.17) are either inhomogeneous or of higher order, one can show that the contraction problem has a solution. One needs to apply again the principle of improving inhomogeneities as shown in detail in Eq. (20.16).

28.3. Connecting the Bulk and the Front

Let us summarize the results we have described so far. For the region $(-\infty, X_\varepsilon]$, we have analyzed in detail the inhomogeneous problem (28.9). We have tacitly assumed that for $|n| \neq 1$, the equation corresponding to (28.2) with an inhomogeneity can be solved in the space H. This equation is of the form

$$(3\varepsilon^2 - (1 - n^2 + 2i\varepsilon n\partial_x + \varepsilon^2\partial_x^2)^2 + \varepsilon^2 c\partial_x)X_n = h_n . \qquad (28.19)$$

We call A_n the operator on the l.h.s. of (28.19). This operator can be analyzed by the "polynomial" methods used earlier and it is easy to see that A_n has a 2 dimensional nullspace in H, namely those functions which decay near $-\infty$. The corresponding eigenfunctions will be called

$$e^{iq_{n,1}x} \text{ and } e^{iq_{n,2}x} ,$$

where $q_{n,j}$, $j = 1, 2$ are the roots of the characteristic polynomial with negative imaginary part. The most general solution of (28.19) is then of the form

$$W_n = A_n^{-1}h - \xi_n e^{iq_{n,1}x} - \zeta_n e^{iq_{n,2}x} ,$$

where A_n^{-1} is a particular inverse, and ξ_n, ζ_n are arbitrary constants (they will have to be related to ξ_{-n}, ζ_{-n} by reality requirements).

We also have nullspaces in the sector $n = 1$. There, we have transformed the problem to a matrix differential equation

$$M\begin{pmatrix} a \\ b \end{pmatrix} = \begin{pmatrix} h_3 \\ h_4 \end{pmatrix} , \qquad (28.20)$$

cf. (28.17). According to (28.18), there are 2 eigenvalues with negative imaginary part, leading to a 2-dimensional nullspace for M. Thus, the general solution of (28.20) is of the form

$$\begin{pmatrix} a \\ b \end{pmatrix} = M^{-1}\begin{pmatrix} h_3 \\ h_4 \end{pmatrix} - \xi_1 v_1 - \zeta_1 v_2 ,$$

with v_1 and v_2 the two eigenvectors with eigenvalue 0. There is one further integration constant for this problem, generated by the translation invariance of the original equation, see (28.12), since

$$L_1 u' = 0$$

and $u' \in H$. Note that $L_2 u$ is also zero, but u is not in H.

We now connect the bulk solution to the front solution. To do this one has to check two things, namely that there are sufficiently many free parameters and that they satisfy a transversality condition. In other words, we want to be sure these parameters are effective. To study this question, one introduces a map S from parameters and amplitudes for the bulk solution to the values of the $W_n(X)$ and their first 3 derivatives.

So we have to find

$$a(x), b(x), W_n(x)|_{n\neq 1}, \xi_n, \zeta_n$$

in such a way that the resulting values at X are in the stable manifold ahead of the front, which we have constructed in Section 28.1. This construction is tedious, but straightforward, and is again an application of the contraction mapping principle.

Outlook

The study of fronts and of instabilities which we have presented here is a subject which is still very much in its infancy, despite large efforts in the scientific community. In this section, we try to convey our general impression on the state of the subject. This account is necessarily very subjective, and may be outdated in the near future. We hope that it indicates the directions of research we consider to be worthwhile.

We choose to illustrate the current difficulties of the subject with one example which we have studied for some time and which is under intensive investigation by a large number of people: the theory of **dendritic growth**. This theory is described in one of the models by a *nonlocal* equation which describes the time evolution of the temperature field $u(x, t)$ when a liquid solidifies. This problem is described by a complicated integro-differential equation; see for example Langer[L]. We call this the dendrite equation. This equation depends on two parameters which are of interest to us, the surface tension in the interface and the anisotropy of the growing solid. When the surface tension is 0, this equation is known to have quasistationary solutions which are parabolas (in the 2-dimensional version) or paraboloids (in the 3-dimensional version), moving with constant speed c. These parabolas are called Ivantsov solutions. As in the case of the real amplitude equation, the speed with which the parabola moves is not determined by the equation. Therefore, we have the

Problem 1. *What is the selection mechanism for quasistationary solutions in problems which have a one parameter family of solutions moving with constant speed?*

We have shown that this problem has a precise answer in the case of the real amplitude equation. In that case, one can appeal to the maximum principle to show that all localized initial conditions tend to the solution with minimal speed; see Section 25. We are not aware of any other argument which leads to a similar result. Langer and Müller-Krumbhaar [LMK] have formulated a marginality conjecture which suggests that in the one-parameter family of solutions the one with critical speed largest basin of attraction. Extrapolating, one can then hope that this large basin of attraction contains a large class of physically relevant initial conditions. This seems like a very reasonable conjecture usually called the marginality conjecture. It somehow says that in phase space, the "last" solution, which is often, but not always, the marginal one, must have a larger basin of attraction than the ones inside the parameter range.

Problem 2. *Try to formalize the above set of ideas.*

In the absence of positive results for the two problems above, researchers became somewhat discouraged and tried to argue that the presence of a one parameter family of solutions is unnatural, and that the original set of equations should be changed. This change is the addition of a surface tension and anisotropy term in the equation. While we do not agree with this reasoning for introducing such a term from the point of view of Problem 1 and Problem 2, it is, of course, very reasonable from a physical point of view to do so. We now analyze what mathematical problems this new, more complicated equation poses. It has been argued, quite convincingly, see [L] for a review of results, that the new equation has stationary solutions only for a *discrete* set of values for the speed. These solutions still look like parabolas. The mathematical argument is a more complicated version of the WKB method described in Section 28. Namely, the perturbation generated by the surface tension is singular with respect to the original operator. Kruskal and Segur [KS] have solved this problem for the dendritic problem above. It would be nice to have an answer to

Problem 3. *Under which conditions does a singular term destroy a continuum of solutions?*

We are now led to ask the next natural question, namely that of *stability analysis* for either the continuous family of solutions or the discrete family of solutions. There is a large number of studies of that question: They are mostly of numerical type, or analysis in perturbation theory. Pillet [P] is working on a study of these problems from a pseudo-differential point of view. All problems encountered in this research seem to be related to

Problem 4. *What is the adequate space of perturbations for the stability analysis of solutions to dendritic problems?*

It would also be interesting to have an answer to the following, more general question:

Problem 5. *What is the general theory of stability of solutions "bifurcated" from a quasistationary solution by singular perturbation?*

That is to say, can one guess directly from the generation of new solutions through the introduction of surface tension and anisotropy what the spectrum of excitations is? It seems to us that it may be easier to find a model equation for Problem 5 than to solve Problem 4.

Assuming all of the problems above are solved, one would then be confronted with actual bifurcations from this parabolic-type quasistationary solutions. Note that there must be some system of parabolic coordinates in this problem. A question related to Problem 4 is then

Problem 6. *What is an adequate definition of fronts for the dendrite equation?*

We end this collection of problems with a question concerning center manifolds. It seems clear to us that the reduction of complicated systems to simpler ones is an important method for the study of time-dependent systems. One cannot expect that every dynamical problem can be reduced to a low-dimensional dynamical system, and one of the aims of this book was to show that results exist even if the effective dynamical problem is infinite dimensional. However, the reduction process is ill understood.

Problem 7. *How does one formulate a center manifold theorem for the case of a continuous spectrum?*

We would like to see a theory in which the use of space-time observation scales helps to overcome the difficulties connected to the continuous spectrum. A related problem is treated, for example, in [BLP].

Notation

space dimension	d
order parameter dimension	ν
time	t
dual time	ω
space	x
dual space	ξ
order parameter	u
physical parameter(s)	α
bifurcation parameter	ε
nonlinear operator	$\mathcal{N}$
nonlinear constraint	$\mathcal{C}$
main branch solution	U
tangent vectors	v
eigenvectors	v_0
linearized operator	$\mathrm{D}\mathcal{N} \equiv \mathrm{A}$
bifurcating solution	u_ε
function space	$\mathbf{X}, \mathbf{Y}$
eigenvalue	σ
nonlinear semigroup	S_t
speed of frame	c
space frequency	ω

Glossary

This book has been written with the intent of explaining the mathematics of extended systems not only to mathematicians, but to non-mathematicians as well. The present glossary re-explains a few important terms in a relatively informal fashion for the reader who might have forgotten one or the other of the definitions. It does not replace any textbook on these subjects and we give further references for the interested reader.

Banach Space. A Banach space is a vector space, equipped with a norm, and complete in the associated metric. That is, every Cauchy sequence converges to an element of the space. The best known Banach spaces in physics are Hilbert spaces. They not only have a norm, but also a scalar product from which the norm derives (i.e. $\|v\| = (v,v)^{1/2}$). In general, we consider infinite dimensional Banach spaces, which are then a generalization of $\mathbf{R}^n$ to infinite n. [DS].

Sobolev Space. A Sobolev space is a Banach space made up (in general) of differentiable functions, with a norm given, for example, by

$$\|f\|^2 = \int dx\, |f(x)|^2 w_0(x) + \int dx\, |f'(x)|^2 w_1(x) ,$$

and w_0, w_1 are non-negative weight functions such as $(1 + x^2)$. The main idea is that the norm contains some sort of bound on first or higher derivatives of f [RS]. In fact, we do not use Sobolev spaces, but rather weighted spaces with the L^∞ norm.

The Contraction Mapping Principle. In vague terms, the contraction mapping principle says that if a (nonlinear) map $\mathcal{N}$ maps a ball of a Banach space $\mathbf{X}$ into itself, and is a contraction, then there is a unique fixed point of $\mathcal{N}$ in this ball. We shall always use the theorem in the following constructive form. Let $\mathcal{B}$ be the ball of radius ρ around zero in X,

$$\mathcal{B} = \{u \in X : \|u\| \le \rho\} .$$

Assume $\|\mathcal{N}(0)\| = \varepsilon$ and, for all $u \in \mathcal{B}$,

$$\|\partial_u \mathcal{N}_u v\| \le \beta \|v\| .$$

If $\varepsilon + \beta\rho < \rho$, then $\mathcal{N}$ maps the ball of radius ρ into itself (strictly) and is a contraction. To see this, it suffices to write

$$\|\mathcal{N}(u_1) - \mathcal{N}(u_2)\| \le \int_0^1 d\alpha\, \|\partial_u \mathcal{N}_{\alpha u_1 + (1-\alpha) u_2} \frac{u_1 - u_2}{\|u_1 - u_2\|}\| .$$

This also means that the fixed point can be found by iterating $\mathcal{N}$ starting from an arbitrary point in the ball. [D].

References

This book is *not* a review of the subject of instabilities, fronts, and the like. The list of references is therefore not a reflection of the activity in the subject. We cite only textbooks and, in addition, a few references which may be needed for an understanding of the book. Good reviews of the subject exist, see for example [Ch], [L], [Man], where many references are given.

[Ar] V. Arnold: *Mathematical Methods of Classical Mechanics*, Berlin, Heidelberg, New York, Springer (1978).

[Ar1] V. Arnold: *Chapitres supplémentaires de la théorie des équations différentielles ordinaires*, Moscow, Mir (1980).

[AVG] V. Arnold, A. Varchenko, and S. Goussein-Zadé: *Singularités des applications différentiables, Part 2*, Mir, Moscow (1986).

[AW] D. Aronson and H. Weinberger: Multidimensional nonlinear diffusion arising in population genetics. Adv. Math. **30**, 33–76 (1978).

[Be] G. R. Belitskii: Equivalence and normal forms of germs of smooth mappings. Russian Math. Surveys **33**, 107–177 (1978).

[BLP] A. Bensoussan, J. L. Lions, and G. Papanicolaou: *Asymptotic analysis for periodic structures*, Amsterdam, North-Holland (1978).

[BPV] P. Bergé, Y. Pomeau, and C. Vidal: *L'ordre dans le chaos*, Paris, Hermann (1984).

[Br] M. Bramson: *Convergence of solutions of the Kolmogorov equation to travelling waves*, Providence, AMS, Memoirs of the AMS **44**, no.285 (1983).

[Ch] S. Chandrasekhar: *Hydrodynamic and Hydromagnetic Stability*, Oxford, Oxford University Press (1961).

[CE1] P. Collet and J.-P. Eckmann: The existence of dendritic fronts. Commun. Math. Phys. **107**, 39–92 (1986).

[CE2] P. Collet and J.-P. Eckmann: The stability of modulated fronts. Helv. Phys. Acta **60**, 969–991 (1987).

[CF] P. Coullet and S. Fauve: Propagative phase dynamics for systems with Galilean invariance. Phys. Rev. Lett. **55**, 2857–2859 (1984).

[CR] M. Crandall and P. Rabinowitz: Bifurcation from simple eigenvalues. J. Funct. Analysis **8**, 321–340 (1971).

[DL] G. Dee and J. Langer: Propagating pattern selection. Phys. Rev. Letter **50**, 383–386 (1983).

[D] J. Dieudonné: *Foundations of Modern Analysis*, New York, London, Academic Press (1968).

[DS] N. Dunford and J. Schwartz: *Linear operators*, New York, Interscience (1958).

[E] W. Eckhaus: *Studies in non-linear stability theory*, Berlin, Heidelberg, New York, Springer, Springer tracts in Nat. Phil. **6** (1965).

[ET] C. Elphick, E. Tirapegui, M. E. Brachet, P. Coullet, and G. Iooss: A simple global characterization for normal forms of singular vector fields. Physica **29D**, 95–127 (1987).

[Fi] P. C. Fife: *Mathematical aspects of reacting and diffusing systems*, Berlin, Heidelberg, New York, Springer, Lecture Notes in Biomathematics **28** (1979).

[F] A. Friedman: *Partial Differential Equations of Parabolic Type*, Prentice-Hall, Englewood Cliffs (1964).

[GS] I. M. Gelfand and G. E. Shilov: *Generalized Functions III*, New York, San Francisco, London, Academic Press (1967).

[GH] J. Guckenheimer and P. Holmes: *Nonlinear Oscillations, Dynamical Systems, and Bifurcations of Vector Fields*, Berlin, Heidelberg, New York, Springer (1983).

[GD] J. D. Gunton and M. Droz: *Introduction to the Theory of Metastable and Unstable States*, Berlin, Heidelberg, New York, Springer, Lecture Notes in Physics **183** (1983).

[HNW] E. Hairer, S. P. Nørsett, and G. Wanner: *Solving Ordinary Differential Equations I*, Berlin, Heidelberg, New York, Springer (1987).

[Ha] P. Hartman: *Ordinary Differential Equations*, New York, Wiley (1964).

[HPS] M. W. Hirsch, C. C. Pugh, and M. Shub: *Invariant Manifolds*, Lecture Notes in Mathematics Vol. 583, Berlin, Heidelberg, New York, Springer (1977).

[H] L. Hörmander: *The Analysis of Linear Partial Differential Equations*, Berlin, Heidelberg, New York, Springer (1983).

[Hu] P. Huerre: Spatio-temporal instabilities in closed and open systems. In *Instabilities and non-equilibrium structures* (E. Tirapegui, D. Villaroel eds). Dordrecht, Boston, Lancaster, Tokyo, Reidel (1987).

[IMD] G. Iooss, A. Mielke, and Y. Demay: Theory of steady Ginzburg-Landau equation, in hydrodynamic stability problems. Preprint Nice University (1988).

[IJ] G. Iooss, D. D. Joseph: *Elementary stability and bifurcation theory*, Berlin, Heidelberg, New York, Springer, Undergraduate Texts In Mathematics (1980).

[K] T. Kato: *Perturbation Theory for Linear Operators*, Berlin, Heidelberg, New York, Springer (1966).

[Ki] K. Kirchgässner: Nonlinear resonant surface waves and homoclinic bifurcation. Adv. Appl. Mech. **26**, 135–181 (1988).

[KS] M. Kruskal and H. Segur: Asymptotics beyond all orders in a model of dendritic crystals. Aero. Res. Ass. of Princeton Tech. Memo (1985).

[La] O. E. Lanford III: *Lectures on Dynamical Systems*. (To appear).

[L] J. S. Langer: Lectures in the Theory of Pattern Formation. In *Chance and matter* (Souletie, Vannimenus, Stora Eds). Amsterdam etc., North-Holland (1987 (Les Houches 1986)).

[LMK] J. S. Langer and H. Müller-Krumbhaar: Mode selection in a dendrite-like nonlinear system. Phys. Rev. **A27**, 499 (1982).

[Man] P. Manneville: *Dissipative Structures and Weak Turbulence*, New York, San Francisco, London, Academic Press (1989).

[M1] V. P. Maslov: *Théorie des perturbations et méthodes asymptotiques*, Paris, Dunod, Gauthier-Villars (1972).

[M2] V. P. Maslov: Asymptotic soliton-form solutions of equations with small dispersion. Russian Math. Surveys **36**, 63–126 (1981).

[M3] V. P. Maslov: Non-standard characteristics in asymptotic problems. Russian Math. Surveys **38**, 3–36 (1983).

[M4] V. P. Maslov: Deterministic theory of turbulence in hydrodynamics (coherent structures). In *VIIIth International Congress of Mathematical Physics* (M. Mebkhout, R. Sénéor eds). Singapore, World Scientific (1987).

[N] A. Newell: The dynamics of patterns: A survey. In *Propagation in Systems far from Equilibrium* (Wesfreid et al. eds). Berlin, Heidelberg, New York, Springer (1988 (Les Houches 1987)).

[P] C.-A. Pillet: (To appear).

[PW] M. H. Protter and H. F. Weinberger: *Maximum Principles in Differential Equations*, Prentice-Hall, Englewood Cliffs (1967).

[RS] M. Reed and B. Simon: *Methods of Modern Mathematical Physics*, New York, San Francisco, London, Academic Press (1972).

[Ro] G. C. Rota: Extension theory of differential operators I. Commun. in Pure and Appl. Math. **11**, 23–65 (1958).

[R1] D. Ruelle: Bifurcations in the presence of a symmetry group. Arch. Rat. Mech. and Analysis **51**, 136–152 (1973).

[R2] D. Ruelle: *Elements of Differentiable Dynamics and Bifurcation Theory*, Boston, Academic Press (1989).

[S] D. Sattinger: Weighted norms for the stability of travelling waves. Journ. Diff. Equ. **25**, 130–144 (1977).

[SB] B. Shraiman and D. Bensimon: Singularities in nonlocal interface dynamics. Phys. Rev. **A30**, 2840–2842 (1984).

[Sh] M. A. Shubin: Almost periodic functions and partial differential operators. Russian Math. Surveys **33**, 1–52 (1978).

[Y] K. Yosida: *Functional Analysis*, Berlin, Heidelberg, New York, Springer (1981).

[ZS] A. Zippelius and E. D. Siggia: Stability of finite-amplitude convection. Phys. Fluids **26**, 2905–2915 (1983).

Analytical Index

The index contains some of technical terms which are used in this book, so that the reader can find their definition or use.

GPSR Authorized Representative: Easy Access System Europe - Mustamäe tee 50, 10621 Tallinn, Estonia, gpsr.requests@easproject.com

www.ingramcontent.com/pod-product-compliance
Ingram Content Group UK Ltd.
Pitfield, Milton Keynes, MK11 3LW, UK
UKHW050735260926
475585UK00004B/21
9780691636177